The Wisdom of God Manifested in the Works of the Creation

By

John Ray
Fellow of the Royal Society

Facsimile edition of the 1826 edition

Published by the Ray Society to mark
the three hundredth anniversary of
John Ray's death in 1705

© The Ray Society

ISBN 0903874326

The Ray Society is a Registered Charity
Number 208082

Sold by: Scion Publishing Ltd,
Bloxham Mill, Barford Road, Bloxham,
Oxfordshire OX15 4FF, UK
www.scionpublishing.com

Printed and Bound by Henry Ling Ltd,
The Dorset Press, 23, High Street East,
Dorchester, DT1 1HD, UK

CONTENTS

Preface vii

Introduction ix

Short Résumé of John Ray's Life xiv

The Wisdom of God, manifested in the
Works of the Creation, by John Ray.
Facsimile of the 1826 Edition 21

PREFACE

The name of John Ray is not as well known as it deserves to be. Not only was he the leading British naturalist of the late seventeenth century, but he was also distinguished as a classical scholar and theologian. His name is known to biologists because of the existence of the Ray Society, founded in the early nineteenth century, which publishes monographs in the field of natural history.

After studying at St. Catherine's College and Trinity College, Cambridge, Ray was elected to a major fellowship at Trinity College in 1660, but forfeited it two years later because his conscience did not allow him to take the oath required under the Act of Uniformity. He and his collaborator Francis Willoughby are commemorated in an impressive pair of busts by Roubiliac, flanking the entrance to the Wren Library of Trinity College. Trinity College also supports the John Ray Trust, set up in 1986. The trust is based in Braintree, Essex, where Ray was at school.

The full title of the book of which this is a facsimile is: *The Wisdom of God Manifested in the Works of the Creation*. Its contents are based on a series of sermons that Ray preached in the chapel of Trinity College on themes such as, the coordination of different parts of the body being so perfect that it could only have been produced by some supernatural agency. This was the only reasonable position to take in Ray's time, nearly 200 years before Darwin produced a natural explanation in the theory of 'the origin of species by means of natural selection'.

March 2005

Andrew Huxley
Trinity College
Cambridge

INTRODUCTION

John Ray's pre-eminence as 'the father of British Natural History' was acknowledged by the founders of the Ray Society in 1844, when they named their new society in his honour. It is therefore fitting that the 300th anniversary of John Ray's death in 1705 should be marked by the Ray Society publishing a facsimile of his most popular *The Wisdom of God Manifested in the Works of the Creation*. It is in this collection of sermons, originally given at Cambridge, that Ray made clear his view that the study of nature was compatible with pleasing God and reconciling the variety clearly observed in his studies of the natural world with the design of a Creator. This work represented a major break with other theologians, who had regarded the natural world as a temptation, leading away from salvation and, consequently, was unsuitable for Christian study. Ray's observation of the natural world and his attempt to place it into a coherent framework, together with his attempt to reconcile his scientific work with his religious belief, led him in *Wisdom of God* to propound the concept of 'natural theology' (physico-theology). In his natural theology, Ray recognised that living things were adapted to their environments and, as such, were fit objects to study in order to glorify God and to understand his purpose in the Creation. Ray studied whole animals and plants and his classification extended to studies on their form and function, based on direct observation. *Wisdom of God* includes many of Ray's observations, made directly on plants and animals. His approach continued through the work of later naturalists and, in particular, those nineteenth century natural historians whose studies and influence culminated in the work of Darwin and Wallace.

The influence of Ray and his interaction with his contemporaries, for example Hooke, Lister and Boyle, as well as his influence on later scientists and theologians, is still a matter of study and debate by historians of scientific thought. However, the point should not be lost that Ray was a man of his times and was embedded in contemporary thought and debate at a time when the study of the natural world was beginning to show a divide between the literal interpretation of the biblical account of creation and what was actually being observed. Ray's ideas and explanation for fossils and the fossil record illustrate the growing dilemma for the natural philosopher of reconciling both the intrinsic process of studying the natural world and the growing *corpus* of knowledge of the natural world, which was at variance with the biblical account of creation. Ray, as a creationist, was disturbed by his conclusion that fossils were the remains of living creatures and his observation that fossils were found in well-marked beds and would consequently have been washed away from land and not onto it by a flood, was not in agreement with the biblical account. It is in this context of reconciling religion with science, that *Wisdom of God* was Ray's most popular book. It was written in English and produced in thirteen editions over a fifty year period, it was also translated into French and German and was extensively plagiarised by other authors. This facsimile of *Wisdom of God*, as well as providing a valuable tool for students studying the development of scientific thought in the seventeenth century also, by allowing a modern audience to access the work, provides an insight into the dilemma of reconciling religious belief with science — a problem of the seventeenth century which still has a resonance today.

An outline of John Ray's life, his rise from humble origins, his father was a blacksmith and his mother a herbalist and healer, to his eminent position as both a theologian and a natural philosopher is given in the short résumé of Ray's life produced with this facsimile. There are a number of more extensive published accounts, in particular, Charles Raven's biography (1950), but also others by subsequent authors. Ray produced numerous works on the natural world starting in 1660 with his *Catalogue of Cambridge Plants* and ending with *Synopsis Methodica Avium et Piscium*, published in 1713 after his death. In his works on the natural world, based on his own direct observations and investigations, Ray attempted to bring order and provide a framework, not merely for naming plants and animals, but also to provide a natural system of classification that he felt would reflect the order imposed by God on his Creation. His whole animal and plant approach was in contrast to Linnaeus, whose main interest was in classification and who used a restricted number of characters. Linnaeus' classification of plants was based on the floral reproductive organs, while Ray, in his *Methodus Plantarum Nova* (1682), used flowers, stems, seeds, fruits and roots and was the first to divide flowering plants into monocotyledons and dicotyledons. Hence he provided the basis of a natural classification for the emerging science of Botany. Although the tenth edition of Linnaeus' *Systema Naturae* is the starting point for all modern taxonomic nomenclature, it is pertinent to note that it was Ray's whole animal and plant approach, with its emphasis on form and function, that influenced the methodological basis for studying the natural world in both the seventeenth and eighteenth centuries and into the nineteenth century.

The Society is indebted to Dr Max Walters* for not only suggesting that it would be fitting that the 300th Anniversary of John Ray's death should be marked by the Ray Society, but also for providing his own copy of the 'Dove Edition of 1826' and allowing it to be unbound to facilitate the production of this facsimile. The Dove Edition is considered to be the last version of *Wisdom of God* that represents Ray's own thoughts; subsequent editions were heavily modified and edited. Dr Walters has also written an essay, published by the Ray Society to accompany this facsimile. This essay, after giving a succinct account of Ray's life, reappraises his relationship to Linnaeus (with particular reference to Ray's biographer Charles Raven's own view on that relationship) and from the starting point of Ray, contrasts the development of botanical with zoological systematics.

August 2005

Nicholas Evans
Honorary Secretary
The Ray Society
c/o The Natural History Museum
London

* Dr Walter's death on the 11th of December 2005 is noted with very deep regret.

SHORT RÉSUMÉ OF RAY'S LIFE

This short résumé is based on the epitome in Charles Raven's biography *John Ray: naturalist* (2nd ed. 1950, Cambridge University Press).

John Wray (he changed the spelling to Ray in 1670) was born on 29 November 1627 at Black Notley, Essex, the third child of the village blacksmith. He was schooled in Braintree, in the Jesus Chapel of St Michael's Church. He went up to Cambridge University in 1644, graduating as a BA in 1648 and by 1649 he was a Minor Fellow at Trinity College, Cambridge. He was gradually promoted within the college over the next twelve years, including appointments as Junior Dean and Steward.

An illness in 1650 led him to start his lifelong study of botany, finding the walks helpful to his convalescence. This led to his first publication in 1660, the *Catalogus Cantabrigiam* (a local flora for Cambridge). In that same year he made his first journey with Francis Willoughby; other journeys with friends around Britain followed during the next two years. He was also ordained deacon and priest that year. However, in 1662 he forfeited his Fellowship, not being willing to take the oath under the Act of Uniformity.

Having to leave Cambridge, Ray commenced a major journey around Europe that covered the years 1663–66. That ended, he made Middleton Hall, the home of his friend Willoughby, his base from 1666 until Willoughby's death in

SHORT RÉSUMÉ OF RAY'S LIFE

1672. During this period he made many excursions around Britain, visiting friends, sometimes in the company of others. In 1667 he was admitted a Fellow of the Royal Society of London. This might have proved a career opportunity for him later when he was sounded out for the Secretaryship in 1677, but he refused it. He published *Catalogus Angliae* (of English plants) and *Collection of English Proverbs*, both in 1670.

He married Margaret Oakley in 1673. In due course they had four daughters, although one died in 1698. He also published, in 1673 *Observations and Catalogus Exteris* (on his continental tour) and *Collection of English Words* (i.e. dialectical ones). In 1675 he published *Dictionariolum trilingue* (i.e. English/Latin/Greek); his later editions of this work were titled *Nomenclator Classicus*. He repaid his debt of honour to Willoughby by substantially editing and publishing in 1676 the latter's *Ornithologia*.

Another significant move occurred in 1679 when his mother died. Earlier he had built 'Dewlands' in Black Notley for her after his father's death. Now he moved into 'Dewlands' himself, ending his itinerant life and marking a shift from exploration to consolidation of his work in many publications. First was *Methodus Plantarum Nova* in 1682. Then, in 1686, came the first volume of his major work *Historia Plantarum*; and also Willoughby's *Historia Piscium*, which he had edited. 1687 marks the first reference to the sores on his legs that seriously affected the health of Ray's later years. They were a painful handicap, even so, publications continued. In 1688 came *Historia Plantarum vol. II* and *Fasciculus Britannicarum* (a supplement towards the Flora). 1690 saw published *Synopsis Britannicarum* (his Flora of Britain).

The, in 1691, Ray published his first edition of **The Wisdom of God Manifested in the Works of the Creation**. A second, enlarged, edition of *Wisdom of God* followed the next year, together with *Miscellaneous Discourses* (much of this was on geology, but from a theological point of view). This was enlarged the next year and re-titled *Three Physico-Theological Discourses*. That same year, 1693, he produced *Synopsis Quadrupedum* and a translation of Rauwolf's *Collection of Curious Travels*. He published *Persuasive to a Holy Life* in 1700, a third edition of *Wisdom of God* in 1701, and *Methodus Emendata* (a botanical update) in 1703. *Historia Plantarum vol. III* was produced in 1704, together with *Methodus Insectorum* and the fourth, and his final, edition of *Wisdom of God*. Ray produced several editions of a number of other works during his life as well, not all of which have been listed in this summary.

On 17 January 1705 John Ray died at 'Dewlands', Black Notley, aged 77. His *Historia Insectorum* and *Synopsis Avium et Piscium* were both published posthumously.

May 2005

Nigel Cooper
Philadelphia House
67 West Street
Isleham, Ely

THE
WISDOM OF GOD,
in the
WORKS OF THE CREATION.

BY JOHN RAY,
Fellow of the Royal Society.

And so leaves the dogs behind her.
Part I.

London:
ENGRAVED FOR DOVE'S ENGLISH CLASSICS.

Corbould. Heath.

Our First Parent saw that the tree and its fruit was pleasant to the eyes, and so was invited to take and eat it.____
Part II.

[DOVE's ENGLISH CLASSICS.]

THE
WISDOM OF GOD

MANIFESTED IN

THE WORKS OF THE CREATION:

IN TWO PARTS.

VIZ.

THE HEAVENLY BODIES, ELEMENTS, METEORS,
FOSSILS, VEGETABLES, ANIMALS
(BEASTS, BIRDS, FISHES, AND INSECTS);

MORE PARTICULARLY

IN THE BODY OF THE EARTH, ITS FIGURE, MOTION,
AND CONSISTENCY,
AND IN THE ADMIRABLE STRUCTURE
OF THE BODIES OF MAN AND OTHER ANIMALS, AS
ALSO IN THEIR GENERATION, &c.

WITH ANSWERS TO SOME OBJECTIONS.

By JOHN RAY,
FELLOW OF THE ROYAL SOCIETY.

LONDON:
PRINTED BY J. F. DOVE;
For the Booksellers of
ENGLAND, SCOTLAND, AND IRELAND.
1826.

JOHN RAY,

The celebrated naturalist, was the son of a blacksmith, at Black Notley, in Essex, and born there, Nov. 29, 1628. He was educated at the school of Braintree, and next at Catherine-hall, Cambridge, but removed from thence to Trinity-college; and in 1649 obtained a fellowship. At the Restoration he was episcopally ordained; but lost his fellowship for refusing to sign the declaration against the solemn league and covenant. After this he devoted himself to the study of nature; and in 1663 accompanied Mr. Willoughby on a tour through Europe; of which journey he published an account in 1673. Mr. Ray was elected a fellow of the Royal Society in 1667; and the transactions of that body afford ample proofs of his attention to science. In 1672 came out his 'Collection of English Proverbs;' previous to which he published his 'Catalogue of English Plants.' The death of Mr. Willoughby was much felt by Mr. Ray, who was appointed one of his executors, and entrusted with the education of his sons. Soon after this he married, and settled at his native place, where he completed his 'Methodus Plantarum Nova,' published in octavo, in 1682; the 'Historia Plantarum,' 3 vols. folio; and the 'Synopsis Methodica Stirpium,' 8vo.

LIFE OF THE AUTHOR.

He also printed Mr. Willoughby's, 'History of Birds;' and that of Fishes. His next publication was entitled 'The Wisdom of God manifested in the Works of the Creation,' which passed through several editions. This was followed by 'Three Discourses concerning the Chaos, Deluge, and Dissolution of the world.' To these pieces succeeded his 'Synopsis Methodica Animalium Quadrupedum;' and in 1693, his 'Sylloge Stirpium Europearum extra Britanniam.' His last undertaking was a 'History of the British Insects,' which he did not live to complete; but his notes, as also his letters, were published after his death, which event happened at Black Notley, Jan. 17, 1705.

TO THE

MUCH HONOURED AND TRULY RELIGIOUS LADY,

THE LADY LETICE WENDY,

OF WENDY IN CAMBRIDGESHIRE.

MADAM,

Two or three reasons induce me to present this discourse to your Ladyship, and to make choice of you for its patroness: first, because I owe it to the liberality of your honoured brother, that I have this leisure to write any thing. Secondly, because also your many and signal favours, seeing I am not in a capacity to requite them, seem to exact from me at least a public acknowledgment, which such a Dedication gives me an opportunity to make. Thirdly, because, of such kind of writings, I know not where to choose a more able judge, or more candid reader. I am sensible that you do so much abhor any thing that looks like flattery, that out of an excess of modesty you cannot patiently bear the hearing of your own just commendations; and therefore, should I enlarge upon that subject, I know I should have but little thanks for my pains.

Indeed, you have much better motives to do well, than the praise of men, the favour of God, peace of conscience, the hope and expectation of a future reward of eternal happiness; and therefore I had rather write of you to others, to provoke them to imitate so excellent an example, than to yourself, to encourage you in your Christian course, and to fortify you in your athletic conflicts with the greatest of temporal evils, bodily pain and anguish; though I do not know, why you should reject any consideration that may conduce to support you under so heavy pressures, and of so long continuance; of which to ingenuous natures, true honour, that is, the concurrent testimony and approbation of good men, is not the meanest. No less man than St. Augustine was doubtful, whether the extremity of bodily pain were not the greatest evil that human nature was capable of suffering: 'Nay (saith he) I was sometimes compelled to consent to Cornelius Celsus, that it was so, neither did his reason seem to me absurd; we being compounded of two parts, soul and body, of which the first is the better, the latter the worser, the greatest good must be the best thing belonging to the better part, that is wisdom; and the greatest evil the worst thing incident to the worser part (the body), that is pain.' Now, though I know not whether this reason be firm and conclusive, yet I am of accord with him, that of all the evils we are sensible of in this world, it is the sorest; the most resolute patience being baffled and prostrated by a fierce and lasting paroxysm of the gout, or stone, or cholic, and compelled to yield to its furious insults, and confess itself vanquished, the soul being unable to divert, or to do any thing else but pore upon the pain. And therefore those stoical vaunts of their wise man's being happy in Perillus's bull, I utterly reject and explode, as vain rhodomontades, and chimerical figments; for that there never was such a wise man among them, nor indeed could be: yet do I not say, that the patience of a good man can be so far conquered by the sharpest and severest torments, as to be compelled to deny or blaspheme God, or his religion; yea, or so much as to complain of his injustice, though perchance

DEDICATION.

he may be brought with Job to curse his day, yet not to curse his God, as his wife tempted him to do.

Now that the great ’Αγωνοθέτης, and Βραβευτὴς, the most just Judge and Rewarder, would be pleased to qualify and mitigate your sufferings as not to exceed the measure of your strength and patience, or else arm you with such a high degree of Christian fortitude, as to be able to grapple with the most extreme, and when you have finished your course in this world, grant you a placid and easy passage out of it, and dignify you as one of his victors, with a crown of eternal glory and felicity, is the prayer of,

Madam,
Your Ladyship's most devoted
in all service,
JOHN RAY.

PREFACE.

In all ages wherein learning hath flourished, complaint hath been made of the itch of writing, and the multitude of worthless books, wherewith importunate scribblers have pestered the world, 'scribimus indocti doctique:' and—'Tenet insanabile multos scribendi cacoëthes.' I am sensible that this tractate may likely incur the censure of a superfluous piece, and myself the blame of giving the reader unnecessary trouble, there having been so much, so well written of this subject, by the most learned men of our time; Dr. Moore, Dr. Cudworth, Dr. Stillingfleet, late bishop of Worcester, Dr. Parker, late of Oxon; and to name no more, the honourable Robert Boyle, Esquire: so that it will need some apology. First, Therefore in excuse of it, I plead, That there are in it some considerations new and untouched by others: wherein, if I be mistaken, I allege, secondly, That the manner of delivery and expression may be more suitable to some men's apprehension, and facile to their understandings. If that will not hold, I pretend, thirdly, That all the particulars contained in this book cannot be found in any one piece known to me, but lie scattered and dispersed in many; and so this may serve to relieve those fastidious readers, that are not willing to take the pains to search them out: and possibly there may be some whose ability (whatever their industry might be) will not serve them to purchase, nor their opportunity to borrow those books, who yet may spare money enough to buy so inconsiderable a trifle. If none of these excuses suffice to acquit me of blame, and remove all prejudice, I have two farther reasons to offer, which I think will reach home, and justify this undertaking. First, That all men who presume to write, at least whose writings the printers will venture to publish, are of some note in the world; and where they do, or have lived and conversed, have some sphere of friends and acquaintance, that know and esteem them, who, it is likely, will buy any book they shall write, for the author's sake, who otherwise would have read none of that subject, though ten times better; and so the book, however inferior to what have been already published, may happen to do much good. Secondly, By virtue of my function, I suspect myself to be obliged to write something in divinity, having written so much on other subjects: for being not permitted to serve the church with my tongue in preaching, I know not but it may be my duty to serve it with my hand in writing; and I have made choice of this subject, as thinking myself best qualified to treat of it. If what I have now written shall find so favourable acceptance, as to encourage me to proceed, God granting life and health, the reader may expect more: if otherwise, I must be content to be laid aside as useless, and satisfy myself in having made this experiment.

As for this discourse, I have been careful to admit nothing for matter of fact, or experiment, but what is undoubtedly true, lest I should build upon a sandy and ruinous foundation; and by the admixture of what is false, render that which is true suspicious.

I might have added many more particulars, nay, my text warrants me to run over all the visible works of God in particular, and to trace the footsteps of his wisdom in the composition, order, harmony, and uses of every one of them, as well as of those that I have selected. But, first, This would be a task far transcending my skill and abilities; nay, the joint skill and endeavours of all men now living, or that shall live after a thousand ages, should the world last so long. 'For no man can find out the work that God maketh from the beginning to the end,' Eccles. iii. 11. Secondly, I was willing to consult the infirmity of the reader, or indeed of mankind in general; which, after a short confinement to one sort of dish, is apt to loathe it, though never so wholesome, and which at first was most pleasant and acceptable: and so to moderate my discourse, as to make an end of writing before I might presume he should be quite tired with reading.

I shall now add a word or two concerning the usefulness of the argument or matter of this discourse, and the reason I had to make choice of it, besides what I have already offered.

First, The belief of a Deity being the foundation of all religion (religion being nothing but a devout worshipping of God, or an inclination of mind to serve and worship him); 'for he that cometh to God, must believe that he is:' it is a matter of the highest concernment, to be firmly settled and established in a full persuasion of this main point: now this must be demonstrated by arguments drawn from the light of nature, and works of creation. For as all other sciences, so divinity proves not, but supposes its subjects, taking it for granted, that by natural light, men are sufficiently convinced of the being of a Deity. There are indeed supernatural demonstrations of this fundamental truth, but not common to all persons or times, and so liable to cavil and exception by atheistical persons, as inward illuminations of mind, a spirit of prophecy and foretelling future contingents, illustrious miracles, and the like. But these proofs, taken from effects and operations exposed to every man's view, not to be denied or questioned by any, are most effectual to convince all that deny or doubt of it. Neither are they only convictive of the greatest and subtlest adversaries, but intelligible also to the meanest capacities. For you may hear illiterate persons of the lowest rank of the commonalty affirming, that they need no proof of the being of a God, for that every pile of grass, or ear of corn, sufficiently proves that: for, say they, all the men of the world cannot make such a thing as one of these; and if they cannot do it, who can, or did make it but God? To tell them, that it made itself, or sprung up by chance, would be as ridiculous as to tell the greatest philosopher so.

Secondly, The particulars of this discourse, serve not only to demonstrate the being of a Deity, but also to illustrate some of his principal attributes; as namely, his infinite power and wisdom. The vast multitude of creatures, and those not only small, but immensely great; the sun and moon, and all the heavenly hosts, are effects and proofs of his almighty power. 'The heavens declare the glory of God, and the firmament sheweth his handy work,' Psal. xix. 1. The admirable contrivance of all and each of them, the adapting all the parts of animals to their several uses: the provision that is made for their sustenance, which is often taken notice of in Scripture, Psal. cxlv. 15, 16. 'The eyes of all wait upon thee, thou givest them their meat in due season. Thou openest thy hand, and satisfiest the desire of every living thing.' Matt. vi. 26. 'Behold the fowls of the air: for they sow not, neither do they reap, nor gather into barns; yet your heavenly Father feedeth them.' Psal. cxlvii. 9.. 'He giveth to the beast his food, and to the young ravens when they cry.' And lastly, Their mutual subserviency to each other, and unanimous conspiring to promote and carry on the public good, are evident demonstrations of his sovereign wisdom.

Lastly, They serve to stir up and increase in us the affections and habits of admiration, humility, and gratitude. Psal. viii. 3. 'When I considered the heavens the work of thy fingers, the moon and the stars which thou hast ordained: what is man that thou art mindful of him, or the son of man that thou visitest him?' And to these purposes the holy Psalmist is very frequent in the enumeration and consideration of these works, which may warrant me doing the like, and justify the denominating such a discourse as this, rather theological than philosophical.

Note, That by the works of the creation, in the title, I mean the works created by God at first, and by him conserved to this day in the same state and condition in which they were at first made; for conservation, according to the judgment both of philosophers and divines, is a continued creation.

CONTENTS.

PART I.

THE multitude of creatures an argument of the wisdom and power of God, 21, 27.

That the fixed stars are innumerable, agreed on all sides, as well by such as embrace the new hypothesis.

That they are as so many suns placed at unequal distances, and each having its planets moving about it, furnished with their inhabitants, like the earth; as by those that adhere to the old, that they are all situate in the same spherical superficies, 21—23.

A guess at the number of terrestrial bodies: 1. Inanimate; as stones, earths, concrete and inconcrete juices, metals, and minerals. 2. Animals; birds, beasts, fishes, and insects. 3. Plants, herbs, and trees, 23—27.

Working the same effect by divers means and instruments, an argument of wisdom. And that God doth this in the works of the Creation, proved by several examples, 27—30.

That the material works of God are wisely contrived and adapted to ends both general and particular, 31.

The Aristotelean hypothesis, That the World was co-eternal with God, condemned, 32.

The Epicurean hypothesis, That the World was made by a casual concurrence and cohesion of atoms, rejected, 32. Their declination of atoms justly derided, and their whole hypothesis ingenuously confuted by Cicero, 34.

The Cartesian hypothesis, That supposing God had only created matter, divided it into a certain number of parts, and put it into motion, according to a few laws, it would of itself have produced the world without any more ado, confuted in Dr. Cudworth's works, 37. 40.

Des Cartes's assertion, That the ends of God in any of his works are equally undiscoverable by us, censured and reproved, 38.

His opinion concerning the cause of the motion of the heart, 42.

The Honourable Mr. Boyle's hypothesis considered, and pleaded against, 46.

The Author's mistake concerning the hypothesis of Mr. Boyle acknowledged, 48.

The parts of the body formed, and situated contrary to the laws of specific gravity, 50, 51.

A plastic nature under God superintending and effecting natural productions, 50.

Their opinion, that hold the souls of brutes to be material, and the whole animal body and soul to be a mere machine, not agreeable to the general sense of mankind, 51.

Of the visible works of God, and their division, 54. The Atomic hypothesis approved, ib.

The works of Nature far more exquisitely formed than the works of art, 55.

The various species of inanimate bodies to be attributed to the divers figures of their principles, or minute component particles, 56. That these principles are naturally indivisible, proved, ib. That they are not very numerous, 57.

Of the heavenly bodies, 57. That the whole universe is divided into two sorts of bodies, viz. thin and fluid; dense and consistent, ib. That this last sort are endued with a two-fold power. 1. Of gravity. 2. Of circular motion, and why, 58. The heavenly bodies moved in the most regular, facile, and convenient manner, 59.

Of the sun, his uses, and the convenience of his situation and motion, 60.

Of the moon, and its uses, 61.

Of the rest of the planets and fixed stars, the regularity and constancy of their motions, whence Cicero rationally infers, that they are governed by reason, 62.

Eclipses useful to settle chronology, and determine longitudes, 63.

Of terrestrial inanimate simple bodies, as elements commonly so called. 1. Fire, its various uses, 64. Of air, its use and necessity for breathing, to all sorts of animals, aquatic as well as terrestrial; nay, in a sort, to plants themselves, 65. The effects and uses of its gravity and elastic quality, 66.

CONTENTS.

That the *fœtus* in the womb hath a kind of respiration, and whence it receives the air, 67.

That the air insinuates itself into water for the respiration of fishes, 69; and even into subterraneous waters, whence it clears the mines of damps, 70. A plastic nature necessary for putting the diaphragm and muscles for respiration into motion at first, 71.

Of water, its uses, 71. Of the sea and its tides, 72. An objection concerning the needlessness of so much sea, of no use to mankind, answered. And the wisdom of God in thus unequally dividing sea and land, manifested and asserted, 72, 73. The use of floods, 74.

That the winds bring up more vapours from the sea, than they carry down thither, 74. 76.

That the motion of the water levels the bottom of the sea, 77. The reason why the sea-plants grow for the most part flat like a fan, and none in the great depths, ib.

Of springs and rivers, baths and mineral waters. Simple water nourisheth not, 78.

Of the earth, its uses and differences, 79.

Of meteors, or bodies imperfectly mixed. And, 1. Of rain, 80. 2. Of wind, and its various uses, 81.

Of inanimate mixed bodies. 1. Stones, their qualities and uses, 82. Particularly of the loadstone, its admirable phenomena, effects, and uses, 85. 2. Metals, their various uses, of great importance to mankind, as iron, without which we could have had nothing of culture or civility: gold and silver for the coining of money, which how many ways useful is shewn out of Dr. Cockburn's Essays, 86.

That the minute component particles of bodies are naturally indivisible, proved, 88, 89.

Of vegetables, or plants, their stature and magnitude, figure, shape, and site of leaves, flowers, and fruits, and their parts all determined, as also their age and duration, 90. The admirable complication of the seminal plant, 91. The uses of the several parts of plants, roots, fibres, vessels, barks, and leaves, 92. The beauty and elegancy of the leaves, flowers, and fruits of plants, 93. That there is such a thing as beauty and comeliness of proportion proved, 94.

The uses of flowers, 95. Of seeds, and their teguments, and observations concerning them, ib. Their lasting vitality or fecundity, 98. The pappous plumage of seeds, 97. The tendrils and prickles of plants, of what use, 99.

That wheat, the best of all grains, is patient both of heat and cold, and scarce refuseth any climate, and that scarce any grain is more fruitful, 99, 100.

Of the signatures of plants, 100.

Of animals, the provision that is made for the continuance of the species, 101. That females have within them, from the beginning, the seeds of all the young they shall ever bring forth, 102. An observation of Cicero's about multiparous creatures, ib. Why birds lay eggs, 103. Of what use the yolk of the egg s to the chicken, 104.

Birds that cannot number, yet omit not any one of their young in feeding of them, 104. Though they cannot number, yet that they can distinguish many from few, proved, ib. The speedy growth of young birds in the nest, 105. The process of building their nests, and incubation, 106. Feeding, breeding, and defending their young, and the admirable Στοργὴ, ib. The due numerical proportion between males and females, in all kinds of animals, kept up constantly, 107. The conveniency of the time of the year, when the several sorts of animals are brought forth, 108. Why birds swallow pebble-stones, 114. The provision of nature for keeping of birds' nests clean, 115. Various strange instincts of animals, 110—114. As that animals should know where their natural weapons are situate, and how to make use of them. That the weak and timorous should be made swift of foot or wing for flight. That they should naturally know their enemies, and such as prey upon them, though they had never seen them before. That as soon as they are brought forth, they should know their proper food. That ducklings, though led by a hen, so soon as they see water, should venture into it, the hen in vain endeavouring to hinder or reclaim them. That birds of the same kind should make nests exactly alike, wheresoever bred, and though they had never seen a nest made.

The migration of birds from one country to another, a strange and unaccountable action, 113.

The wonderful art observable in the construction, situation, and figure of the cells in honey-combs, 116. That bees, and other animals, lay up in store, either for the food of their young, or their own winter-provision, 117.

The provision that is made for the preservation and security of weak and timorous creatures, and for the diminishing of the rapacious, 119.

The fitness of the parts of the bodies of animals to every one's own nature and manner of living, instanced in, 1. The swine, 122, &c. 2. The mole, 123. 3. The tamandua, or ant-bear, 124. 4. The cameleon, 125. 5. The whole genius of woodpeckers, ib. 6. Swallows, 126. 7. Doukers or loons, ib.

In birds, all the members are fitted for the use of flying, 127. The use of the tail in birds, 128.

The bodies of fishes most conveniently formed and provided for the use of swimming, 131. And particularly those of cetaceous fishes for respiration, and preserving their natural heat, 132. And of amphibious creatures, 133.

The fitting of the parts of animals one to another, viz. The genitals of the sexes, 133. The nipples of the paps to the mouth and organs of suction, ib. The admirable structure of the breasts or paps, for the preparing and separating, the containing and retaining of the milk, that it doth not flow out without pressure or suction. ib.

Several observations of Aristotle's, relating to the fitness of the parts to the creature's nature and manner of living, and to their respective uses, 134, &c.

Another remarkable instance in proportioning the length of the neck to that of legs in animals, 136. Of the aponeurosis in the neck, why given to most quadrupeds, and not to man, 137. That some birds have but short legs, and yet long necks, and why, ib. That this instance cannot be accounted for by atheists, 138.

The various kinds of voices the same animal uses on divers occasions, and to divers purposes, argumentative of providence and counsel, in conferring them upon it, being so extremely useful and serviceable to the creature, 138.

An objection against the uses of several bodies I have instanced in, relating to man, answered, 139. A discourse in the person of Almighty God to man, 140, &c.

The incredible smallness and subtleness of some animalcules, an argument and proof of the admirable art of the Creator, 144.

Practical inferences from the precedent discourse,

wherein it is shewn, That the world was in some sense made for man, yet not so as to have no other end or use of its creation, but to serve him, 146. The contemplation and consideration of the works of God, may probably be some part of our employment in heaven, 148.

The sun, moon, stars, &c. are called upon to praise, which they can do no otherwise, than by affording man, and other intelligent beings, matter or subject of praising him. And therefore, men and angels are called upon to consider those great effects of the divine power, wisdom, and goodness, and to give God the praise and glory due to him for them, 154. That God doth, and may justly intend and design his own glory, 157.

CONTENTS OF PART II.

I. OF the whole body of the earth: and first of its figure, which is demonstrated to be spherical, 160. The conveniences of this figure shewn for union of parts, strength, convenience of habitation, and circular motion, upon its own poles, 161.

II. Of its motion, both diurnal upon its own poles, and annual in the ecliptic; and both these shewn to be rational, and not dissonant to Scripture, 163, &c. The present direction, and constant parallelism of its axis to itself, shewn to be most convenient for the inhabitants of the earth, by the inconvenience of any different position, 165, &c. That the torrid zone is habitable, and stored with great multitudes of men, and other animals, contrary to the opinion of some of the ancients, 167. Neither are the heats there prejudicial to the longevity of men, 168. A digression to prove, that the lives of men are longest in the hottest countries, 169. That it would not be more convenient for the inhabitants of the earth, if the tropics stood at a greater distance, proved, 171. A very considerable and heretofore unobserved convenience of this inclination of the earth's axis, which Mr. Keill affords us, 170.

Of the usefulness of the present figure, constitution, and consistency of the several parts of this terraqueous globe, 172.

An enumeration of some plants, which afford almost whatever is necessary for human life, 174, &c.

CONTENTS. xix

Plants having a kind of cisterns, or basins, formed of their leaves compacted together, for the containing and preserving of water during the dry seasons of the year, for their own nourishment, and for the relief and refreshment of birds, insects, and even men themselves, 175, 176.

Of mountains, and their uses, 180—184.

II. The wisdom of God discovered in the structure of the bodies of man, and other animals.

Eleven general observations, demonstrating this wisdom and goodness of God in forming our bodies.

1. The erect posture of the body of man, 184. Commodious, (1.) For the sustaining of the head, 185. (2. For prospect, ib. (3.) For walking and using our hands, 186. That this erection of the body was intended by nature, proved by several arguments, and particularly by the fastening the cone of the heart to the midriff, of which an account is given out of Dr. Tyson's anatomy of the ourang outang, 187.

2. In that there is nothing deficient or redundant in the body of man, 189. A notable story of a man's giving suck to a young child, well attested, ib.

3. The parts of the body most conveniently situate, for use, for ornament, and for mutual assistance, 190.

4. Ample provision made for the defence and security of the principal parts, heart, brain, and lungs, 192.

5. Abundant provision made against evil accidents and inconveniences, 194. Some observations concerning sleep, 196.

That the unsensibleness of pain during sleep proceeds rather from the relaxation of the nerves than their obstruction, 197.

6. The constancy observed in the number, figure, and site of the principal parts, and the variety in the less, 198.

7. The annexing of pleasure to those actions that are necessary for the support of the individuum, and the continuation and propagation of the kind, 199.

8. The multitude of intentions the Creator must have in the formation, and fitting the several parts for their respective actions and uses, 199.

9. The fitting and accommodating some parts to divers offices and uses, an argument of wisdom and design in the contrivance of the body of man, and other animals, 200.

xx   CONTENTS.

10. In the nourishment of their bodies, making that food which is proper to preserve them in a healthful state, grateful to the taste, and agreeable to the stomach. Of the great use of pain, 201.

11. The variety of lineaments and dissimilitude of the faces of men, as also of their voices, and hand-writings, all of mighty importance to man, 203.

Of the particular parts of the body: and 1. Of the head and hair, 206. The reason of baldness. 2. Of the eye: its beauty, ib. Its humours and tunicles transparent, 207. (1.) For the clearness. (2.) For the distinctness of vision, 208. The parts of the eye of a figure most convenient for the collection of the visual rays, viz. convex, ib. The uveous tunicle hath a musculous power for contracting and dilating of the pupil, 209. Its inside, and that of the choriods, why blackened, ib. The figure of the eye alterable, according to the exigency of the object, in respect of distance or propinquity, 210. Why the optic nerve is not inserted right behind the eye, 211. Why, though the rays be decussated in the pupil of the eye, the object is not seen inverted, ib. The use of the aqueous humour, and that it is suddenly reparable, 212. The Tunica Cornea protuberant above the white of the eye, and why, 213. The use of the muscles of the eye, 214. The provision for the defence and security of this precious part, ib. The uses of the eye-lids, and their frequent winking, 215. That as man wants, so he needs not the seventh or suspensory muscle, which is of great use and necessity to brutes, 216. The need and use of the nictating membrane in brutes, and that man needs it not, 217.

Thirdly, Of the ear, 217. The use of the auricula, 218. Of the tympanum of the ear, its bones, and their muscles, and of the use of the ear-wax, &c. ib.

Fourthly, Of the teeth, nine remarkable observations concerning their situation, structure, and uses, 219.

Fifthly, of the tongue, and its various uses, for tasting and gathering of food, for managing of mastication, for forming of words, 222. Speaking proper to man, ib. Of the Ductus Salivales, and of the great use of the saliva, or spittle, 223.

Sixthly, Of the windpipe, its admirable structure and uses, 223.

CONTENTS.

Seventhly, Of the heart, the use and necessity of its pulse for the circulation of the blood, and the admirable make and contrivance of it for that office, 225. Of the muscular coat, and pulse of the arteries, effected by a kind of constriction, or peristaltic motion, and not merely by a wave of the blood every pulse, 226.

The wonderful artifice of nature in regulating the motion of the blood in the veins and arteries, by assisting and promoting it in the one, and moderating it in the other, 228. The use of the vena sine pari, 230.

Eighthly, of the hand, its structure and various uses, not easily to be enumerated, 230.

Ninthly of the back-bone, its figure, and why divided into vertebres, 233.

The provision that is made for the easy and expedite motion of the bones in their articulations by a two-fold juice: 1. An oily one supplied by the marrow. 2. Mucilaginous, prepared and separated by certain glandules made for that purpose, 234. This inunction of the head of the bones, with these juices, is useful, 1. To facilitate motion. 2. To prevent incalescency. 3. To prevent attrition, 235.

Why the bones are made hollow, 237.

Why the belly is fleshy, and not enclosed with bones like the breast, 238.

The motion of the guts, ib. Of the liver and use of the gall, 239.

Of the bladder, its structure and use. Of the kidneys and glandules and ureters, their composition and uses, 239.

The adapting all the bones, muscles, and vessels, to their several uses, and the joining and compacting of them together noted, 239.

The geometrical contrivance of the muscles, and fitting them for their several motions and actions, according to the exactest rules of mechanics, 240.

The packing and thrusting together such a multitude of various and different parts so close, that there should be no unnecessary vacuity in the body, nor any clashing between them, but mutual assistance, admirable, 240.

Membranes capable of a prodigious extension, use, in gestation of twins, &c., 240.

The parts that seem of little or no use, as the fat, shewn

to be greatly useful, 241. How separated from the blood, and received into it again, 242.

The consideration of the formation of the fœtus in the womb waived, and why, 243.

What a fitness the seed hath to fashion and form, and why the child resembles the parents, and sometimes the ancestor, 244.

The construction of a set of temporary parts, for the use of the fœtus only while in the womb, a clear proof of design, 245.

No equivocal or spontaneous generation, but that all animals are generated by animal parents of their own kind, 246. And probably all plants too produced by seed, and none spontaneous, proved and vindicated, and the objections against it answered, 248—268.

That the cossus of the ancients was not the hexapod of a beetle, as I thought, but an eruca, agreed with Dr. Lister, 254.

The louse searching out sordid and nasty clothes to harbour and breed in, probably designed to deter men and women from sluttishness and uncleanliness, 255.

An additional and most effectual argument against spontaneous generation; viz. That there are no new species of animals produced, 256.

Whence those vast numbers of small frogs, which have been observed to appear upon refreshing showers, after drought, do probably proceed, shewn in an instance of his own observation, by Mr. Derham, 260.

Of toads found in the heart of timber trees, and in the middle of great stones, 266.

Miscellaneous observations concerning the structure, actions, and uses of some parts of animals omitted in the first part: as also the reasons of some instincts and actions of brutes, 269. The swine's snout fitted for digging up of roots, which are his natural food, as likewise the porpoise for rooting up of sand-eels, 122.

The manner and organs of respiration accommodated to the temper of animals; their place and manner of living shewn in three sorts of respiration. 1. By lungs, with two ventricles of the heart in hotter animals, 269. 2. By lungs with but one ventricle. 3. By gills with one only ventricle of the heart, 271, &c. Why the foramen ovale is kept open in some amphibious animals, 272. In some of them the epiglottis is large, and why, 273. No epiglottis in elephants, and why; and how that creature secures himself from mice creeping up his proboscis into his lungs, ib.

Two notable observations of the sagacity of the tortoise, the one of the land, the other of the sea tortoise, 274, 275.

The armour of the hedgehog, and tatou, and their power of contracting themselves into a round ball, a great instance of design for their defence and security, 276.

The manner and use of the extending and withdrawing the curtain of the periophthalmium, or nictating membrane, in beasts and birds, 278. That the aqueous humour of the eye will not freeze, 282.

Of the make of a camel's foot, and his bags to reserve water in his stomach for his needs, 283.

The use of rapacious creatures swallowing some of the hair, fur, and feathers, of the beasts or birds they prey upon, 284.

A conjecture by what means cartilaginous fishes raise and sink themselves in the water, 285.

That Nature employs all the methods and artifices of chymists, in analyzing of bodies, and separating their parts, and outdoes them, and the several particulars instanced in, 286.

Observations about the gullet and diaphragm, 287.

An admirable story out of Galen, about the taking a kid out of the womb of its dam, and bringing it up by hand, and remarks upon it, 287, &c.

The natural texture of membranes so made, as to be immensely dilatable, of great use and necessity in gestation, 291.

A notable instance of providence, in the make of the veins and arteries near the heart, 292.

An answer to an objection against the wisdom of God, in making inferior ranks of creatures, 301.

The Atheist's main subterfuge and pretence, to elude and evade all our arguments and instances, to demonstrate the necessity of providence, design, and wisdom, in the formation of all parts of the world, viz. That things made uses, and not uses things, precluded and confuted, 294, &c.

Of the use of those vast numbers of prodigiously small insects that are bred in the waters, 307.

An objection against the wisdom of God, in creating such a multitude of useless insects, and some also noxious and pernicious to man, and other animals, answered, and the various uses of them declared, 303.

Many practical inferences and observations, from 309 to the end of the book.

THE WISDOM OF GOD

IN THE CREATION.

PART I.

How manifold are thy works, O Lord! In wisdom hast thou made them all.—PSAL. civ. 24.

In these words are two clauses, in the first whereof the Psalmist admires the multitude of God's works, 'How manifold are thy works, O Lord!' In the second he celebrates his wisdom in the creation of them; 'In wisdom hast thou made them all.'

Of the first of these I shall say little, only briefly run over the works of this visible world, and give some guess at the number of them; whence it will appear, that upon this account they well deserve admiration, the number of them being uninvestigable by us, and so affording us a demonstrative proof of the unlimited extent of the Creator's skill, and the fecundity of his wisdom and power. That the number of corporeal creatures is unmeasurably great, and known only to the Creator himself, may thus probably be collected: first of all, the number of fixed stars is on all hands acknowledged to be next to infinite: secondly, every fixed star, in the now-received hypothesis, is a sun or sun-like body, and in like manner incircled with a chorus of planets moving about it; for the fixed stars are not all placed in one and the same concave spherical superficies, and equidistant from us, as they seem to be, but are variously and disorderly situate, some nearer, some farther off, just like trees in a wood or forest; as Gassendus exemplifies them. And as in a wood, though the trees grow never so irregularly, yet the eye of the spectator, wherever placed, or whithersoever removed, describes still a circle of trees: so would it in like manner wherever it were in the forest of stars, describe a spherical superficies about it. Thirdly, each of these planets is in all likelihood furnished with as great variety of corporeal creatures, animate and inanimate, as the earth is, and all as different in nature as they are in place from the terrestial, and from each other. Whence it will follow that these must be much more infinite than the stars: I do not mean absolutely, according to philosophic exactness, infinite; but only infinite or innumerable as to us, or their number prodigiously great.

That the fixed stars are innumerable, may thus be made out: those visible to the naked eye are by the least account acknowledged to be above a thousand, excluding those towards the south pole, which are not visible in our horizon: besides these, there have been incomparably more detected and brought to light by the telescope; the milky way being found to be (as was formerly conjectured) nothing but great companies or swarms of minute stars singly invisible, but by reason of their proximity mingling and confounding their lights, and appearing like lucid clouds. And it is likely that, had we more perfect telescopes, many thousands more might be discovered; and yet, after all, an incredible multitude remain, by reason of their immense distance, beyond all ken by the best telescopes that could possibly be invented or polished by the wit and hand of an angel. For if the world be, as Des Cartes would have it, indefinitely extended; that is, so far as no human intellect can fancy any bounds of it; then what we see, or can come to see, must be the least part of what is undiscoverable by us, the whole universe extending a thousand times farther beyond the utmost stars we can possibly descry, than those be distant from the earth we live upon. This hypothesis of the fixed stars being so many suns, &c. seems more agreeable to the divine greatness and magnificence. But that which induces me much to doubt of the magnitude of the universe, and immense distance of the fixed stars, is the stupendous phenomena of comets, their sudden accension or appearance in full magnitude, the length of their tails and swiftness of their motion, and gradual diminution of bulk and motion, till at last they disappear. That the universe is indefinitely extended, Des Cartes upon a false ground (that the formal ratio of a body was nothing but extension into length, breadth, and profundity, or having *partes extra partes*, and that body and space were synonymous terms) asserted: it may as well be limited this way, as in the old hypothesis, which places the fixed stars in the same spherical superficies; according to which (old hypothesis) they may also be demonstrated by the same mediums to be innumerable, only instead of their distance, substituting their smallness for the reason of their invisibility.

But leaving the celestial bodies, I come now to the terrestrial; which are either inanimate or animate. The inanimate are the elements, meteors, and fossils, of all sorts, at the number of which last I cannot give any probable guess: but if the rule, which some considerate philosophers deliver, holds good, viz. how much more imperfect any genus or order of beings is, so much more numerous are the species contained under it: as for example; birds being a more perfect kind of animals than fishes, there are more of these than of those; and for the like reason more birds than quadrupeds, and more insects than of any of the rest, and so more plants than animals, nature being more sparing in her more excellent productions. If this rule, I say, holds good, then should there be more species of fossils, or generally of inanimate bodies, than of vegetables; of which there is some reason to doubt, unless we will admit all sorts of formed stones to be distinct species.

Animate bodies are divided into four great genera or orders, beasts, birds, fishes, and insects.

The species of beasts, including also serpents, are not very numerous; of such as are certainly known and described, I dare say not above 150. And yet I believe not many, that are of any considerable bigness, in the known regions of the world, have escaped the cognizance of the curious. [I reckon all dogs to be of one species, they mingling together in generation, and the breed of such mixtures being prolific.]

The number of birds known and described may be near 500; and the number of fishes, secluding shell-fish, as many; but if the shell-fish be taken in, more than six times the number. How many of each genus remain yet undiscovered, one cannot certainly nor very nearly conjecture; but we may suppose the whole sum of beasts and birds to exceed by a third part, and fishes by one half, those known.

The insects, if we take in the exanguious both terrestrial and aquatic, may, in derogation to the precedent rule, for number, vie even with plants themselves: for the exanguious alone, by what that learned and critical naturalist, my honoured friend, Dr. Martin Lister, hath already observed and delineated, I conjecture, cannot be fewer than 3000 species, perhaps many more.

The butterflies and beetles are such numerous tribes, that I believe in our own native country alone the species of each kind may amount to 150 or more. And if we should make the caterpillars and hexapods, from whence these come, to be distinct species, as most naturalists have done, the number will be doubled, and these two genera will afford us 600 species. But if those be admitted for distinct species, I see no reason but their aureliæ also may pretend to a specific difference from the caterpillars and butterflies, and so we shall have 300 species more; therefore we exclude both these from the degree of species, making them to be the same insect under a different larva or habit.

The fly-kind, if under that name we comprehend all other flying insects, as well such as have four, as such as have but two wings, of both which kinds there are many subordinate genera, will be found in multitude of species to equal, if not exceed, both the forementioned kinds.

The creeping insects that never come to be winged, though for number they may fall short of the flying or winged, yet are they also very numerous; as by running over the several kinds I could easily demonstrate. Supposing then there be a thousand several sorts of insects in this island and the sea near it, if the same proportion holds between the insects native of England, and those of the rest of the world, as doth between plants domestic and exotic (that is, as I guess, near a decuple), the species of insects in the whole earth (land and water) will amount to 10,000, and I do believe they rather exceed than fall short of that sum. Since the writing hereof, having this summer, ann. 1691, with some diligence prosecuted the history of our English insects, and making collections of the several species of each tribe, but particularly and especially of the butterflies, both nocturnal and diurnal, I find the number of such of these alone as breed in our neighbourhood (about Braintree and Noteley in Essex) to exceed the sum I last year assigned to all England, having myself observed and described about 200 kinds great and small, many yet remaining, as I have good reason to believe, by me undiscovered. This I have, since the writing hereof, found true in experience, having every year observed not a few new kinds: nor do I think that, if I should live twenty years longer, I should by my utmost diligence and industry in searching them out, come to an end of them. If then, within a small compass of a mile or two, there are so many species to be found, surely the most modest conjecture cannot estimate the number of all the kinds of papilios native of this island to fall short of 300, which is twice so many as I last summer guessed them to be. Wherefore, using the same argumentations, the number of all the British insects will amount to 2000, and the total sum of those of the whole earth will be 20,000. The number of plants contained in C. Bauhin's Pinax, is about 6000, which are all that had been described by the authors that wrote before him, or observed by himself; in which work, besides mistakes and repetitions incident to the most wary and knowing men in such a work as that, there are a great many, I might say some hundreds, put down for different species, which in my opinion are but accidental varieties: which I do not say to detract from the excellent pains and performance of that learned, judicious, and laborious herbarist, or to defraud him of his deserved honour, but only to shew, that he was too much swayed by the opinions then generally current among herbarists, that different colour or multiplicity of leaves in the flower, and the like accidents, were sufficient to constitute a specific difference. But supposing there had been 6000 then known and described, I cannot think but that there are in the world more than triple that number; there being in the vast continent of America as great a variety of species as with us, and yet but few common to Europe, or perhaps Africa and Asia. And if, on the other side the equator, there be much land still remaining undiscovered, as probably there may, we must suppose the number of plants to be far greater.

What can we infer from all this? if the number of creatures be so exceeding great, how great, nay, immense, must needs be the power and wisdom of him who formed them all! For (that I may borrow the words of a noble and excellent author) as it argues and manifests more skill by far in an artificer, to be able to frame both clocks and watches, and pumps, and mills, and granadoes, and rockets, than he could display in making but one of those sorts of engines; so the Almighty discovers more of his wisdom in forming such a vast multitude of different sorts of creatures, and all with admirable and irreprovable art, than if he had created but a few; for this declares the greatness and unbounded capacity of his understanding. Again: the same superiority of knowledge would be displayed, by contriving engines of the same kind, or for the same purposes, after different fashions, as the moving of clocks or other engines by springs instead of weights; so the infinitely wise Creator hath shewn in many instances, that he is not confined to one only instrument for the working one effect, but can perform the same thing by divers means. So, though feathers seem necessary for flying, yet hath he enabled several creatures to fly without them, as two sort of fishes, one sort of lizard, and the bat, not to mention the numerous tribes of flying insects. In like manner, though the air-bladder in fishes seems necessary for swimming, yet some are so formed as to swim without it, *viz.* First the cartilagineous kind, which by what artifice they poise themselves, ascend and descend at pleasure, and continue in what depth of water they list, is as yet unknown to us. Secondly, the cetaceous kind, or sea-beasts, differing in nothing almost from quadrupeds but the want of feet. The air which in respiration these receive into their lungs, may serve to render their bodies equiponderant to the water; and the constriction or dilatation of it, by the help of the diaphragm and muscles of respiration, may probably assist them to ascend or descend in the water, by a light impulse thereof with their fins.

Again: Though the water being a cold element, the most wise God hath so attempered the blood and bodies of fishes in general, that a small degree of heat is sufficient to preserve their due consistency and motion, and to maintain life; yet to shew that he can preserve a creature in the sea, and in the coldest part of the sea too, that may have as great a degree of heat as quadrupeds themselves, he hath created variety of these cetaceous fishes, which converse chiefly in the Northern seas, whose whole body being encompassed round with a copious fat or blubber (which, by reflecting and redoubling the internal heat, and keeping off the external cold, doth the same thing to them that clothes do to us), is enabled to abide the greatest cold of the sea water. The reason why these fishes delight to frequent chiefly the Northern seas, is, I conceive, not only for the quiet which they enjoy there, but because the Northern air, which they breathe in, being more fully charged with those particles supposed nitrous, which are the aliment of fire, is fittest to maintain the vital heat in that activity which is sufficient to move such an unwieldy bulk as their bodies are with celerity, and to bear up against and repel the ambient cold; and may likewise enable them to continue longer under water than a warmer and a thinner air could.

Another instance to prove that God can and doth by different means produce the same effect, is the various ways of extracting the nutritious juice out of the aliment, in several kinds of creatures.

1. In man and viviparous quadrupeds, the food moistened with the spittle (saliva) is first chewed and prepared in the mouth, then swallowed into the stomach, where being mingled with some dissolvent juices, it is by the heat hereof concocted, macerated, and reduced into a chyle or cremor, and so evacuated into the intestines, where, being mixed with the choler and pancreatic juice, it is farther subtilized and rendered so fluid and penetrant, that the thinner and finer part of it easily finds its way in at the straight orifices of the lacteous veins.

2. In birds there is no mastication or comminution of the meat in the mouth; but in such as are not carnivorous, it is immediately swallowed into the crop or craw, or at least into a

kind of ante-stomach (which I have observed in many, especially piscivorous birds), where it is moistened and mollified by some proper juice from the glandules distilling in there, and thence transferred into the gizzard or musculous stomach, where by the working of the muscles compounding the sides of that ventricle, and by the assistance of small pebbles (which the creature swallows for that purpose), it is, as it were, by millstones ground small, and so transmitted to the guts, to be farther attenuated and subtilized by the forementioned choler and pancreatic juice.

3. In oviparous quadrupeds, as chamelions, lizards, frogs, as also in all sorts of serpents, there is no mastication or comminution of the meat, either in mouth or stomach; but as they swallow insects or other animals whole, so they avoid their skins unbroken, having a heat, or spirits, powerful enough to extract the juice they have need of, without breaking that which contains it; as the Parisian academists tell us. I myself cannot warrant the truth of the observation in all. Here, by the by, we take notice of the wonderful dilatability or extensiveness of the throats and gullets of serpents: I myself have taken two entire adult mice out of the stomach of an adder, whose neck was not bigger than my little finger. These creatures, I say, draw out the juice of what they swallow without any comminution, or so much as breaking the skin; even as it is seen that the juice of grapes is drawn as well from the rape,* where they remain whole, as from a vat, where they are bruised; to borrow the Parisian philosophers' similitude.

4. Fishes, which neither chew their meat in their mouths, nor grind it in their stomachs, do

---

* Whole grapes plucked from the cluster, and wine poured upon them in a vessel.

by the help of a dissolvent liquor, there by nature provided, corrode and reduce it, skin, bones, and all, into a chylus or cremor; and yet (which may seem wonderful) this liquor manifests nothing of acidity to the taste : but notwithstanding how mild and gentle soever it seems to be, it corrodes flesh very strangely and gradually, as aquafortis or the like corrosive waters do metals, as appears to the eye; for I have observed fish in the stomachs of others thus partially corroded; first the superficial part of the flesh, and then deeper and deeper by degrees to the bones.

I come now to the second part of the words, 'In wisdom hast thou made them all;' in discoursing whereof I shall endeavour to make out in particulars what the Psalmist here asserts in general concerning the works of God, that they are all very wisely contrived and adapted to ends both particular and general.

But before I enter upon this task, I shall, by way of preface or introduction, say something concerning those systems which undertake to give an account of the formation of the universe by mechanical hypotheses of matter, moved either uncertainly, or according to some catholic laws, without the intervention and assistance of any superior immaterial agent.

There is no greater, at least no more palpable and convincing argument of the existence of a Deity, than the admirable art and wisdom that discovers itself in the make and constitution, the order and disposition, the ends and uses of all the parts and members of this stately fabric of heaven and earth; for if in the works of art, as for example, a curious edifice or machine, counsel, design, and direction to an end appearing in the whole frame, and in all the several pieces of it, do necessarily infer the being and operation of some intelligent architect or engineer, why shall not also in the works of nature, that grandeur and magnificence, that excellent contrivance for beauty, order, use, &c. which is observable in them, wherein they do as much transcend the effects of human art as infinite power and wisdom exceeds finite, infer the existence and efficacy of an omnipotent and all-wise Creator?

To evade the force of this argument, and to give some account of the original of the world, atheistical persons have set up two hypotheses.

The first is that of Aristotle, that the world was from eternity in the same condition that now it is, having run through the successions of infinite generations; to which they add, self-existent and unproduced. For Aristotle doth not deny God to be the efficient cause of the world; but only asserts, that he created it from eternity, making him a necessary cause thereof; it proceeding from him by way of emanation, as light from the sun.

This hypothesis, which hath some shew of reason, for something must necessarily exist of itself; and if something, why may not all things? This hypothesis, I say, is so clearly and fully confuted by the reverend and learned Dr. Tillotson, late lord archbishop of Canterbury, and primate of all England, in his first printed sermon, and the right reverend father in God, John, late lord bishop of Chester, in Book I. chap. v. of his 'Treatise of the Principles of Natural Religion,' that nothing material can by me be added; to whom therefore I refer the reader.

The Epicurean hypothesis rejected.

The second hypothesis is that of the Epicureans, who held, that there were two principles

IN THE CREATION. 33

self-existent, First, Space, or vacuity; Secondly, Matter or body; both of infinite duration and extension. In this infinite space or vacuity, which hath neither beginning, nor end, nor middle; no limits or extremes, innumerable minute bodies, into which the matter was divided, called atoms, because by reason of their perfect solidity they were really indivisible (for they hold no body capable of division, but what hath vacuities interspersed with matter), of various but a determinate number of figures, and equally ponderous, do perpendicularly descend, and by their fortuitous concourse make compound bodies, and at last the world itself. But now, because if all these atoms should descend plumb down with equal velocity, as according to their doctrine they ought to do, being (as we said) all perfectly solid and imporous, and the vacuum not resisting their motion, they would never the one overtake the other, but, like the drops of a shower, would always keep the same distances, and so there could be no concourse or cohesion of them, and consequently nothing created; partly to avoid this destructive consequence, and partly to give some account of the freedom of will (which they did assert contrary to the Democritic fate) they did absurdly feign a declination of some of these principles, without any shadow or pretence of reason. The former of these motives you have set down by Lucretius, de Nat. Rerum, l. 2. in these words:

> Corpora, cum deorsum rectum per inane feruntur
> Ponderibus propriis, incerto tempore forte,
> Incertisque locis, Spatio discedere paulum;
> Tantum quod momen mutatum dicere possis.

And again:

> Quod, nisi declinare solerent, omnia deorsum,
> Imbris uti guttæ, caderent per inane profundum
> Nec foret offensus natus, nec plaga creata
> Principiis, ita nil unquam natura creâsset.

> Now seeds in downward motion must decline,
> Though vary little from the exactest line,
> For did they still move strait, they needs must fall
> Like drops of rain, dissolv'd and scatter'd all,
> For ever tumbling through the mighty space,
> And never join to make one single mass.

The second motive they had to introduce this gratuitous declination of atoms, the same poet gives us in these verses, lib. 2.

> ——Si semper motus connectitur omnis,
> Et veteri exoritur semper novus ordine certo;
> Nec declinando faciunt primordia motus
> Principium quoddam, quod fati fœdera rumpat,
> Ex infinito ne causam causa sequatur;
> Libera per terras unde hæc animantibus exstat,
> Unde est hæc, inquam, fatis avolsa, voluntas?

> Besides, did all things move in direct line,
> And still one motion to another join
> In certain order, and no seeds decline,
> And make a motion fit to dissipate
> The well-wrought chain of causes and strong fate;
> Whence comes that freedom living creatures find?
> Whence comes the will so free, so unconfin'd,
> Above the power of the fate?

The folly and unreasonableness of this ridiculous and ungrounded figment, I cannot better display and reprove than in the words of Cicero, in the beginning of his first book *de finibus Bonorum et Malorum*. This declination, saith he, is altogether childishly feigned, and yet neither doth it at all solve the difficulty, or effect what they desire. For first they say the atoms decline, and yet assign no reason why. Now nothing is more shameful and unworthy a natural philosopher ('turpius physico') than to assert any thing to be done without a cause, or to give no reason of it. Besides, this is contrary to their own hypothesis taken from sense, that all weights do naturally move perpendicularly downward. Secondly, Again supposing this were true, and that there were such a declination of atoms, yet will it not effect what they intend; for either they do all decline, and so there will be no more concourse than if they did perpendicularly descend; or some decline, and some fall plumb down, which is ridiculously to assign distinct offices and tasks to the atoms, which are all of the same nature and solidity. Again, in his book *de Fato,* he smartly derides this fond conceit thus: What cause is there in nature which turns the atoms aside? Or do they cast lots among themselves which shall decline, which not? Or why do they decline the least interval that may be, and not a greater? Why not two or three *minima* as well as one? 'Optare hoc quidem est, non disputare.' For neither is the atom by any extrinsical impulse diverted from its natural course; neither can there be any cause imagined in the vacuity through which it is carried, why it should not move directly; neither is there any change made in the atom itself, that it should not retain the motion natural to it, by force of its weight or gravity.

As for the whole atomical hypothesis, either Epicurean or Democritic, I shall not, nor need I, spend time to confute it; this having been already solidly and sufficiently done by many learned men, but especially Dr. Cudworth, in his 'Intellectual System of the Universe,' and the present bishop of Worcester, Dr. Stillingfleet in his *Origines Sacræ*. Only I cannot omit the Ciceronian confutation thereof, which I find in the place first quoted, and in his first and second books *de Natura Deorum,* because it may serve as a general introduction to the following particulars. Such a turbulent concourse of atoms could never, saith he, 'hunc mundi ornatum efficere,' compose so well-ordered and beautiful a structure as the world; which therefore both in Greek and Latin hath from thence ('ab ornatu et munditie') obtained its name. And again most fully and appositely in his second *de Nat. Deorum.*

If the works of nature are better, more exact and perfect than the works of art, and art effects nothing without reason; neither can the works of nature be thought to be effected without reason. For, is it not absurd and incongruous, that when thou beholdest a statue or curious picture, thou shouldest acknowledge that art was used to the making of it; or when thou seest the course of a ship upon the waters, thou shouldst not doubt but the motion of it is regulated and directed by reason and art; or when thou considerest a sun-dial or clock, thou shouldst understand presently, that the hours are shewn by art, and not by chance; and yet imagine or believe, that the world, which comprehends all these arts and artificers, was made without counsel or reason? If one should carry into Scythia or Britain such a sphere as our friend Possidonius lately made, each of whose conversions did the same thing in the sun and moon and other five planets, which we see effected every night and day in the heavens, who among those barbarians would doubt that that sphere was composed by reason and art? A wonder then it must needs be, that there should be any man found so stupid and forsaken of reason, as to persuade himself, that this most beautiful and adorned world was or could be produced by the fortuitous concourse of atoms. He that can prevail with himself to believe this, I do not see why he may not as well admit, that if there were made innumerable figures of the one and twenty letters, in gold, suppose, or any other metal, and these well shaken and mixed together, and thrown down from some high place to the ground, they when they lighted upon the earth would be so disposed and ranked, that a man might see and read in them Ennius's Annals; whereas it were a great chance if he should find one verse thereof among them all. For if this concourse of atoms could make a whole world, why may it not sometimes make, and why hath it not somewhere or other in the earth made a temple, or a gallery, or a portico, or a house, or a city? which yet it is so far from doing, and every man so far from believing, that should any one of us be cast, suppose, upon a desolate island, and find there a magnificent palace, artificially contrived according to the exactest rules of architecture, and curiously adorned and furnished, it would never once enter into his head, that this was done by an earthquake, or the fortuitous shuffling together of its component materials; or that it had stood there ever since the construction of the world, or first cohesion of atoms; but would presently conclude that there had been some intelligent architect there, the effect of whose art and skill it was. Or should he find there but upon one single sheet of parchment or paper an epistle or oration written, full of profound sense, expressed in proper and significant words, illustrated and adorned with elegant phrase; it were beyond the possibility of the wit of man to persuade him that this was done by the temerarious dashes of an unguided pen, or by the rude scattering of ink upon the paper, or by the lucky projection of so many letters at all adventures; but he would be convinced by the evidence of the thing at first sight, that there had been not only some man, but some scholar there.

The Cartesian hypothesis considered and censured.

Having rejected this atheistic hypothesis of Epicurus and Democritus, I should now proceed to give particular instances of the art and wisdom clearly appearing in the several parts and members of the universe; from which we may justly infer this general conclusion of the Psalmist, 'In wisdom hast thou made them all:' but that there is a sort of professed theists, I mean Monsieur Des Cartes and his followers, who endeavour to disarm us of this decretory weapon, to evacuate and exterminate this argument, which hath been so successful in all ages to demonstrate the existence, and enforce the belief of a Deity; and to convince and silence all atheistic gainsayers. And this they do,

First, by excluding and banishing all considerations of final causes from natural philosophy; upon pretence, that they are all and every one in particular undiscoverable by us; and that it is rashness and arrogance in us to think we can find out God's ends, and be partakers of his counsels. 'Atque ob hanc unicam rationem totum illud causarum genus, quod a fine peti solet, in rebus physicis nullum usum habere existimo; non enim absque temeritate me puto investigare posse fines Dei.' Medit. Metaph. 'And for this only reason, I think, all that kind of causes which is wont to be taken from the end, to have no use in physics or natural matters; for I cannot without rashness think myself able to find out the ends of God.' And again, in his principles of philosophy: 'Nullas unquam rationes circa res naturales a fine quem Deus aut natura in iis faciendis sibi proposuit admittimus, quia non tantum nobis debemus arrogare, ut ejus conciliorum participes esse possimus.' 'We can by no means admit any reasons about natural things taken from the end which God or nature proposed to themselves in making of them: because we ought not to arrogate so much to ourselves, as to think we may be partakers of his councils.' And more expressly in his fourth answer, viz. to Gassendus's Objections: 'Nec fingi potest, aliquos Dei fines magis quam alios in propatulo esse: omnes enim in imperscrutabili ejus sapientia abysso sunt eodem modo reconditi;' that is, neither can or ought we to feign or imagine that some of God's ends are more manifest than others; for all lie in like manner or equally hidden in the unsearchable abyss of his wisdom.

This confident assertion of Des Cartes is fully examined and reproved by that honourable and excellent person Mr. Boyle, in his 'Disquisition about the final Causes of Natural Things,' sect. i. from page 10. to the end; and therefore I shall not need to say much to it, only in brief this, that it seems to me false and of evil consequence, as being derogatory from the glory of God, and destructive of the acknowledgment and belief of a Deity:

For first, seeing, for instance, that the eye is employed by man and all animals for the use of vision, which, as they are framed, is so necessary for them, that they could not live without it; and God Almighty knew that it would be so; and seeing it is so admirably fitted and adapted to this use, that all the wit and art of men and angels could not have contrived it better, if so well, it must needs be highly absurd and unreasonable to affirm, either that it was not designed at all for this use, or that it is impossible for man to know whether it was or not.

Secondly, how can man give thanks and praise to God for the use of his limbs and senses, and those his good creatures which serve for his sustenance, when he cannot be sure they were made in any respect for him; nay, when it is as likely they were not, and that he doth but abuse them to serve ends for which they were never intended.

Thirdly, this opinion, as I hinted before, supersedes and cassates the best medium we have to demonstrate the being of a Deity, leaving us no other demonstrative proof but that taken from the innate idea; which, if it be a demonstration, is but an obscure one, not satisfying many of the learned themselves, and being too subtile and metaphysical to be apprehended by vulgar capacities, and consequently of no force to persuade and convince them.

Secondly, they endeavour to evacuate and disannul our great argument, by pretending to solve all the phenomena of nature, and to give an account of the production and efformation of the universe, and all the corporeal beings therein, both celestial and terrestrial, as well animate as inanimate, not excluding animals themselves, by a slight hypothesis of matter so and so divided and moved. The hypothesis you have in Des Cartes's 'Principles of Philosophy,' part ii. all the matter of this visible world is by him supposed 'to have been at first divided by God into parts nearly equal to each other, of a mean size, viz. about the bigness of those whereof the heavenly bodies are now compounded; all together having as much motion as is now found in the world; and these to have been equally moved severally every one by itself about its own centre, and among one another, so as to compose a fluid body; and also many of them jointly, or in company, about several other points so far distant from one another, and in the same manner disposed as the centres of the fixed stars now are.' So that God had no more to do than to create the matter, divide it into parts, and put it into motion according to some few laws, and that would of itself produce the world and all the creatures therein.

For a confutation of this hypothesis, I might refer the reader to Dr. Cudworth's system, p. 603, 604; but for his ease I will transcribe the words:—' God in the mean time standing by as an idle spectator of this *lusus atomorum*, this sportful dance of atoms, and of the various results thereof. Nay, these mechanic theists have here quite outstripped and outdone the atomic atheists themselves, they being much more extravagant than ever those were; for the professed atheists durst never venture to affirm, that this regular system of things resulted from the fortuitous motions of atoms at the very first, before they had for a long time together produced many other inept combinations, or aggregate forms of particular things, and nonsensical systems of the whole, and they supposed also that the irregularity of things here in this world would not always continue such neither, but that some time or other confusion and disorder will break in again. Moreover, that besides this world of ours, there are at this very instant innumerable other worlds irregular, and that there is but one of a thousand or ten thousand among the infinite worlds that have such regularity in them; the reason of all which is, because it was generally taken for granted, and looked upon as a common notion, that τῶν ἀπὸ τύχης καὶ τοῦ αὐτομάτου οὐδὲν ἀεὶ οὕτω γίνεται, as Aristotle expresseth it; none of those things which are from fortune or chance come to pass always alike. But our mechanic theists will have their atoms never so much as once to have fumbled in these their motions, nor to have produced any inept system, or incongruous forms at all, but from the very first all along to have taken up their places, and ranged themselves so orderly, methodically, and directly, as that they could not possibly have done it better had they been directed by the most perfect wisdom. Wherefore these atomic theists utterly evacuate that grand argument for a God, taken from the phenomenon of the artificial frame of things, which hath been so much insisted upon in all ages, and which commonly makes the strongest impression of any other upon the minds of men, &c. the atheists in the mean time laughing in their sleeves, and not a little triumphing to see the cause of theism thus betrayed by its professed friends and assertors, and the grand argument for the same totally slurred by them, and so their work done, as it were, to their hands.

Now as this argues the greatest insensibility of mind, or sottishness and stupidity in pretended theists, not to take the least notice of the regular and artificial frame of things, or of the signatures of the divine art and wisdom in them, nor to look upon the world and things of nature with any other eyes than oxen and horses do ; so are there many phenomena in nature, which being partly above the force of these mechanic powers, and partly contrary to the same, can therefore never be salved by them, nor without final causes and some vital principle : as for example, that of gravity or the tendency of bodies downward, the motion of the diaphragm in respiration, the systole and diastole of the heart, which is nothing but a muscular constriction and relaxation, and therefore not mechanical, but vital. We might also add, among many others, the intersection of the planes of the equator and ecliptic, or the earth's diurnal motion upon an axis not parallel to that of the ecliptic, nor perpendicular to the plane thereof : for though Des Cartes would needs imagine this earth of ours once to have been a sun, and so itself the centre of a lesser vortex, whose axis was then directed after this manner, and which therefore still kept the same site or posture, by reason of the striate particles finding no fit pores or traces for their passages through it, but only in this direction; yet does he himself confess, that because these two motions of the earth, the annual and diurnal, would be much more conveniently made upon parallel axes, therefore, according to the laws of mechanism they should be perpetually brought nearer and nearer together, till at length the equator and ecliptic come to have their axis parallel, which as it has not come to pass, so neither hath there been for the last two thousand years (according to the best observations and judgments of astronomers) any nearer approach made of them one to another. Wherefore the continuation of these two motions of the earth, the annual and diurnal, upon axes not parallel is resolvable into nothing but a final and mental cause, or the τὸ Βέλτιστον, because it was best it should be so, the variety of the seasons of the year depending thereupon. But the greatest of all the particular phenomena, is the formation and organization of the bodies of animals, consisting of such variety and curiosity, that these mechanic philosophers being no way able to give an account thereof from the necessary motion of matter, 'unguided by mind for ends,' therefore prudently break off their system there, when they should come to animals, and so leave it altogether untouched. We acknowledge indeed there is a posthumous piece extant, imputed to Cartes, and entituled, *De la formation du Fœtus*, wherein there is some pretence made to salve all this fortuitous mechanism. But as the theory thereof is built wholly upon a false supposition, sufficiently confuted by our Harvey in his book of Generation, 'that the seed doth materially enter into the composition of the egg;' so is it all along precarious and exceptionable: nor doth it extend at all to differences that are in several animals, nor offer the least reason why an animal of one species might not be formed out of the seed of another. Thus far the Doctor, with whom for the main I do consent. I shall only add, that natural philosophers, when they endeavour to give an account of any of the works of nature by preconceived principles of their own, are for the most part grossly mistaken and confuted by experience; as Des Cartes in a matter that lay before him, obvious to sense, and infinitely more easy to find out the cause of, than to give an account of the formation of the world; that is, the pulse of the heart, which he attributes to an ebulition and sudden expansion of the blood in its ventricles, after the manner of the milk, which being heated to such a degree, doth suddenly, and as it were all at once, flush up and run over the vessel. Whether this ebullition be caused by a nitro-sulphureous ferment lodged especially in the left ventricle of the heart, which mingling with the blood excites such an ebullition, as we see made by the mixture of some chymical liquors, viz. oil of vitriol, and deliquated salt of tartar; or by the vital flame warming and boiling the blood. But this conceit of his is contrary both to reason and experience: for, first, it is altogether unreasonable to imagine and affirm, that the cool venal blood should be heated to so high a degree in so short a time as the interval of two pulses, which is less than the sixth part of a minute. Secondly, in cold animals, as for example, eels, the heart will beat for many hours after it is taken out of the body, yea, though the ventricle be opened, and all the blood squeezed out. Thirdly, the process of the fibres which compound the sides of the ventricles running in spiral lines from the tip to the base of the heart, some one way, and some the contrary, do clearly shew that the systole of the heart is nothing but a muscular constriction, as a purse is shut by drawing the strings contrary ways: which is also confirmed by experience; for if the vertex of the heart be cut off, and a finger thrust up into one of the ventricles, in every systole the finger will be sensibly and manifestly pinched by the sides of the ventricle. But for a full confutation of this fancy, I refer the reader to Dr. Lower's 'Treatise de Corde,' Chap. 2. And Des Cartes's rules concerning the transferring of motion from one body in motion to another in motion or in rest, are the most of them by experience found to be false; as they affirm who have made trial of them.

This pulse of the heart Dr. Cudworth would have to be no mechanical, but a vital motion, which to me seems probable, because it is not under the command of the will, nor are we conscious of any power to cause or to restrain it; but it is carried on and continued without our knowledge or notice, neither can it be caused by the impulse of any external movent, unless it be heat. But how can the spirits agitated by heat, unguided by a vital principle, produce such a regular reciprocal motion? If that site which the heart and its fibres have in the diastole be most natural to them (as it seems to be), why doth it again contract itself, and not rest in that posture? If it be once contracted in a systole by the influx of the spirits, why, the spirits continually flowing in without let, doth it not always remain so? [for the systole seems to resemble the forcible bending of a spring, and the diastole its flying out again to its natural site.] What is the spring and principal efficient of this reciprocation? What directs and moderates the motions of the spirits? They being but stupid and senseless matter, cannot of themselves continue any regular and constant motion, without the guidance and regulation of some intelligent being. You will say, what agent is it which you would have to effect this. The sensitive soul it cannot be, because that is indivisible, but the heart, when separated wholly from the body in some animals, continues still to pulse for a considerable time; nay, when it hath quite ceased, it may be brought to beat anew by the application of warm spittle, or by pricking it gently with a pin or needle. I answer, it may be in these instances, the scattering spirits remaining in the heart, may for a time, being agitated by heat, cause these faint pulsations; though I should rather attribute them to a plastic nature or vital principle, as the vegetation of plants must also be.

But, to proceed, neither can I wholly acquiesce in the hypothesis of that honourable and deservedly famous author, I formerly had occasion to mention, which I find in his 'Free Inquiry into the Vulgar Notion of Nature,' pp. 77, 78. delivered in these words: 'I think it probable, that the great and wise Author of things did, when he first formed the universe and undistinguished matter into the world, put its parts into various motions, whereby they were necessarily divided into numberless portions of differing bulks, figures, and situations, in respect of each other: and that by his infinite wisdom and power he did so guide and overrule the motions of these parts, at the beginning of things, as that (whether in a shorter or a longer time reason cannot determine) they were finally disposed into that beautiful and orderly frame that we call the world; among whose parts some were so curiously contrived, as to be fit to become the seeds or seminal principles of plants and animals. And I farther conceive, that he settled such laws or rules of local motion among the parts of the universal matter, that by his ordinary and preserving concourse, the several parts of the universe, thus once completed, should be able to maintain the great construction or system and economy of the mundane bodies, and propagate the species of living creatures.' The same hypothesis he repeats again, pp. 124, 125. of the same treatise.

This hypothesis, I say, I cannot fully acquiesce in, because an intelligent being seems to me requisite to execute the laws of motion. For first, motion being a fluent thing, and one part of its duration being absolutely independent upon another, it doth not follow that because any thing moves this moment, it must necessarily continue to do so the next; unless it were actually possessed of its future motion, which is a contradiction; but it stands in as much need of an efficient to preserve and continue its motion as it did at first to produce it. Secondly, Let matter be divided into the subtilest parts imaginable, and these be moved as swiftly as you will, it is but a senseless and stupid being still, and makes no nearer approach to sense, perception, or vital energy than it had before; and do but only stop the internal motion of its parts, and reduce them to rest, the finest and most subtile body that is may become as gross, and heavy, and stiff as steel or stone. And as for any external laws or established rules of motion, the stupid matter is not capable of observing or taking any notice of them, but would be as sullen as the mountain was that Mahomet commanded to come down to him, neither can those laws execute themselves. Therefore there must, besides matter and law, be some efficient, and that either a quality or power inherent in the matter itself, which is hard to conceive, or some external and intelligent agent, either God himself immediately, or some plastic nature.

Happening lately to read 'The Christian Virtuoso,' written by the same author of the 'Inquiry into the Vulgar Notions of Nature,' the illustrious Mr. Boyle, I find therein these words: 'Nor will the force of all that has been said for God's special providence be eluded by saying with some deists, That after the first formation of the universe all things are brought to pass by the settled laws of nature. For though this be confidently, and not without colour, pretended, yet I confess it doth not satisfy me.—— For I look upon a law as a moral not physical cause, as being indeed but a notional thing, according to which an intelligent and free agent is bound to regulate its actions. But inanimate bodies are utterly incapable of understanding what it is, or what it enjoins, or when they act conformably or unconformably to it. Therefore the actions of inanimate bodies, which cannot incite or moderate their own actions, are produced by real power, not by laws.'

All this being consonant to what I have here written, against what I took to be this honourable person's hypothesis, I must needs, to do him right, acknowledge myself mistaken; perceiving now, that his opinion was, that God Almighty did not only establish laws and rules of local motion among the parts of the universal matter, but did and does also himself execute them, or move the parts of matter, according to them. So that we are in the main agreed, differing chiefly about the agent that executes those laws, which he holds to be God himself immediately, we a plastic nature; for the reasons alleged by Dr. Cudworth, in his 'System,' p. 149. which are, First, Because the former, according to vulgar apprehension, would render the divine Providence operose, solicitous, and distractious: and thereby make the belief of it entertained with greater difficulty, and give advantage to atheists. Secondly, It is not so decorous in respect of God, that he should αὐτουργεῖν ἅπαντα, set his own hand, as it were, to every work, and immediately do all the meanest and triflingest things himself drudgingly, without making use of any inferior or subordinate minister. These two reasons are plausible, but not cogent; the two following are of greater force. Thirdly, The slow and gradual process that is in the generation of things, which would seem to be a vain and idle pomp or trifling formality, if the agent were omnipotent. Fourthly, Those ἁμαρτήματα, as Aristotle calls them, those errors and bungles which are committed when the matter is inept or contumacious, as in monsters, &c. which argue the agent not to be irresistible; and that nature is such a thing as is not altogether incapable, as well as human art, of being sometimes frustrated and disappointed by the indisposition of the matter: whereas an omnipotent agent would always do its work infallibly and irresistibly, no ineptitude or stubbornness of the matter being ever able to hinder such a one, or make him bungle or fumble in any thing. So far the Doctor. For my part, I should make no scruple to attribute the formation of plants, their growth, and nutrition, to the vegetative soul in them; and likewise the formation of animals to the vegetative power of their souls; but that the segments and cuttings of some plants, nay, the very chips and smallest fragments of their body, branches, or roots, will grow and become perfect plants themselves, and so the vegetative soul, if that were the architect, would be divisible, and consequently no spiritual or intelligent being; which the plastic principle must be, as we have shewn. For that must preside over the whole economy of the plant, and be one single agent, which takes care of the bulk and figure of the whole, and the situation, figure, texture of all the parts, root, stalk, branches, leaves, flowers, fruit, and all their vessels and juices. I therefore incline to Dr. Cudworth's opinion, that God uses for these effects the subordinate ministry of some inferior plastic nature; as in his works of providence he doth of angels. For the description whereof I refer the reader to his 'System.'

Secondly, In particular I am difficult to believe, that the bodies of animals can be formed by matter divided and moved by what laws you will or can imagine, without the immediate presidency, direction, and regulation of some intelligent being. In the generation or first formation of, suppose, the human body, out of (though not an homogenous liquor, yet) a fluid substance, the only material agent or mover is a moderate heat. Now how this, by producing an intestine motion in the particles of the matter, which can be conceived to differ in nothing else but figure, magnitude, and gravity, should by virtue thereof, not only separate the heterogeneous parts, but assemble the homogeneous into masses or systems, and that not each kind into one mass, but into many and disjoined ones, as it were so many troops; and that in each troop the particular particles should take their places, and cast themselves into such a figure; as for example, the bones being about three hundred, are formed of various sizes and shapes, so situate and connected, as to be subservient to many hundred intentions and uses, and many of them conspire to one and the same action, and all this contrarily to the laws of specific gravity, in whatever posture the body be formed; for the bones, whose component parts are the heavier, will be above some parts of the flesh, which are the lighter; how much more then, seeing it is formed with the head (which for its bigness is the heaviest of all the parts) uppermost. This, I say, I cannot by any means conceive. I might instance in all the homogeneous parts of the body, their sites and their figures, and ask by what imaginable laws of motion their bulk, figure, situation, and connexion can be made out? What account can be given of the valves, of the veins and arteries of the heart, and of the veins elsewhere, and of their situation; of the figure and consistency of all the humours and membranes of the eye, all conspiring and exactly fitted to the use of seeing? But I have touched upon that already, and shall discourse of it largely afterward. You will ask me, Who or what is the operator in the formation of the bodies of man and other animals? I answer, The sensitive soul itself, if it be a spiritual and immaterial substance, as I am inclinable to believe: but if it be material, and consequently the whole animal but a mere machine or automaton, as I can hardly admit, then must we have recourse to a plastic nature.

That the soul of brutes is material, and the whole animal, soul and body, but a mere machine, is the opinion publicly owned and declared of Des Cartes, Gassendus, Dr. Willis, and others; the same is also necessarily consequent upon the doctrine of the Peripatetics, viz. that the sensitive soul is educed out of the power of the matter, for nothing can be educed out of the matter, but what was there before, which must be either matter or some modification of it. And therefore they cannot grant it to be a spiritual substance, unless they will assert it to be educed out of nothing. This opinion, I say, I can hardly digest. I should rather think animals to be endowed with a lower degree of reason, than that they are mere machines. I could instance in many actions of brutes that are hardly to be accounted for without reason and argumentation; as that commonly noted of dogs, that running before their masters, they will stop at a divarication of the way, till they see which hand their masters will take; and that when they have gotten a prey, which they fear their masters will take from them, they will run away and hide it, and afterward return to it. What account can be given why a dog being to leap upon a table, which he sees to be too high for him to reach at once, if a stool or chair happens to stand near it, doth first mount up that, and from thence the table? If he were a machine or piece of clockwork, and this motion caused by the striking of a spring, there is no reason imaginable why the spring being set on work, should not carry the machine in a right line toward the object that put it in motion, as well when the table is high as when it is low: whereas I have often observed the first leap the creature hath taken up the stool, not to be directly towards the table, but in a line oblique and much declining from the object that moved it, or that part of the table on which it stood.

Many the like actions there are, which I shall not spend time to relate. Should this be true, that beasts were automata or machines, they could have no sense or perception of pleasure or pain, and consequently no cruelty could be exercised towards them; which is contrary to the doleful significations they make when beaten or tormented, and contrary to the common sense of mankind, all men naturally pitying them, as apprehending them to have such a sense and feeling of pain and misery as themselves have; whereas no man is troubled to see a plant torn, or cut, or stamped, or mangled how you please; and at least seemingly contrary to the Scripture too, for it is said, Prov. xii. 10. 'A righteous man regardeth the life of his beast; but the tender mercies of the wicked are cruel.' The former clause is usually Englished, 'A good man is merciful to his beast;' which is the true exposition of it, as appears by the opposite clause, that 'the wicked are cruel.' What less then can be inferred from this place, than that cruelty may be exercised towards beasts, which were they mere machines, it could not be? To which I do not see what can be answered, but that the Scripture accommodates itself to the common, though false opinion of mankind, who take these animals to be endued with sense of pain, and think that cruelty may be exercised towards them; though in reality there is no such thing. Besides, having the same members and organs of sense as we have, it is very probable they have the same sensations and perceptions with us. To this Des Cartes answers, or indeed saith, he hath nothing to answer; but that if they think as well as we, they have an immortal soul as well as we: which is not at all likely, because there is no reason to believe it of some animals without believing it of all; whereas there are many too imperfect to believe it of them, such as are oysters and sponges, and the like. To which I answer, that there is no necessity they should be immortal, because it is possible they may be destroyed or annihilated. But I shall not wade farther into this controversy, because it is beside my scope, and there hath been as much written of it already as I have to say, by Dr. More, Dr.

Cudworth, Des Cartes, Dr. Willis, and others, *pro* and *con.*

Of the visible Works of God, and their Division.

I come now to take a view of the works of the creation, and to observe something of the wisdom of God discernible in the formation of them; in their order and harmony, and in their ends and uses. And first I shall run them over slightly, remarking chiefly what is obvious and exposed to the eyes and notice of the more careless and incurious observer. Secondly, I shall select one or two particular pieces, and take a more exact survey of them; though even in these more will escape our notice than can be discovered by the most diligent scrutiny; for our eyes and senses, however armed or assisted, are too gross to discern the curiosity of the workmanship of nature, or those minute parts by which it acts, and of which bodies are composed; and our understanding too dark and infirm to discover and comprehend all the ends and uses to which the infinitely wise Creator did design them.

But before I proceed, being put in mind thereof by the mention of the assistance of our eyes, I cannot omit one general observation concerning the curiosity of the works of nature in comparison of the works of art, which I shall propose in the late bishop of Chester's words, Treat. of Nat. Religion, lib. i. c. 6. 'The observations which have been made in these latter times by the help of the microscope, since we had the use and improvement of it, discover a vast difference between natural and artificial things. Whatever is natural, beheld through that, appears exquisitely formed, and adorned with all imaginable elegancy and beauty. There are such inimitable gildings in the smallest seeds of plants, but especially in the parts of animals, in the head or eye of a small fly; such accuracy, order, and symmetry in the frame of the most minute creatures, a louse, for example, or a mite, as no man were able to conceive without seeing of them. Whereas the most curious works of art, the sharpest and finest needle doth appear as a blunt rough bar of iron, coming from the furnace or the forge: the most accurate engravings or embossments seem such rude, bungling, and deformed work, as if they had been done with a mattock or a trowel; so vast a difference is there betwixt the skill of nature, and the rudeness and imperfection of art. I might add, that the works of nature, the better lights and glasses you use, the more clearer and exactly formed they appear; whereas the effects of human art, the more curiously they are viewed and examined, the more of deformity they discover.'

This being premised, for our more clear and distinct proceeding in our cursory view of the creation, I shall rank the parts of this material and visible world under several heads. Bodies are either inanimate or animate. Inanimate bodies are either celestial or terrestrial: celestial, as the sun, moon, and stars: terrestial are either simple as the four elements, fire, water, earth, and air; or mixt, either imperfectly, as the meteors, or more perfectly, as stones, metals, minerals, and the like. Animate bodies are either such as are endued with a vegetative soul, as plants; or a sensitive soul, as the bodies of animals, birds, beasts, fishes, and insects; or a rational soul, as the body of man and the vehicles of angels, if any such there be.

I make use of this division to comply with the common and received opinion, and for easier comprehension and memory; though I do not think it agreeable to philosophic verity and accuracy, but do rather incline to the atomick hypothesis: for these bodies we call elements are not the only ingredients of mixed bodies; neither are they absolutely simple themselves, as they do exist in the world, the sea-water containing a copious salt manifest to sense; and both sea and fresh-water sufficing to nourish many species of fish, and consequently containing the various parts of which their bodies are compounded. And I believe there are many species of bodies which the Peripatetics call mixed, which are as simple as the elements themselves, as metals, salts, and some sorts of stones. I should therefore, with Dr. Grew and others, rather attribute the various species of inanimate bodies to the divers figures of the minute particles of which they are made up: and the reason why there is a set and constant number of them in the world, none destroyed, nor any new ones produced, I take to be, because the sum of the figures of those minute bodies into which matter was at first divided, is determinate and fixed. 2. Because those minute parts are indivisible, not absolutely, but by any natural force; so that there neither is nor can be more or fewer of them: for were they divisible into small and diversely-figured parts by fire or any other natural agent, the species of nature must be confounded, some might be lost and destroyed, but new ones would certainly be produced; unless we could suppose these new diminutive particles should again assemble and marshal themselves into corpuscles of such figures as they compounded before; which I see no possibility for them to do, without some Θεὸς ἀπὸ μηχανῆς to direct them: not that I think these inanimate bodies to consist wholly of one sort of atoms, but that their bulk consists mainly or chiefly of one sort. But whereas it may be objected, that metals (which of all others seem to be most simple) may be transmuted one into another, and so the species doth not depend upon the being compounded of atoms of one figure; I answer, I am not fully satisfied of the matter of fact: but if any such transmutation be, possibly all metals may be of one species, and the diversity may proceed from the admixture of different bodies with the principles of the metal. If it be asked, why may not atoms of different species concur to the composition of bodies? and so, though there be but a few sorts of original principles, may there not be produced infinite species of compound bodies, as by the various dispositions and combinations of twenty-four letters innumerable words may be made up? I answer, because the heterogeneous atoms or principles are not naturally apt to cohere and stick together when they are mingled in the same liquor, as the homogeneous readily do.

I do not believe that the species of principles or indivisible particles are exceeding numerous; but possibly the immediate component particles of the bodies of plants and animals may be themselves compounded.

Of the heavenly bodies.

Before I come to treat of the heavenly bodies in particular, I shall premise in general, that the whole universe is divided into two sorts of bodies, the one very thin and fluid, the other more dense, solid, and consistent. The thin and fluid is the ether, comprehending the air or atmosphere encompassing the particular stars and planets. Now, for the stability and perpetuity of the whole universe, the divine wisdom and providence hath given to the solid and stable parts a two-fold power, one of gravity, and the other of circular motion. By the first they are preserved from dissolution and dissipation, which the second would otherwise infer. For it being by the consent of philosophers an innate property of every body moved circularly about any centre to recede, or endeavour to recede from that centre of its motion, and the more strongly the swifter it is moved, the stars and planets being whirled about with great velocity, would suddenly, did nothing inhibit it, at least in a short time, be shattered in pieces, and scattered every way through the ether; but now their gravity unites and binds them up fast, hindering the dispersion of their parts. I will not dispute what gravity is; only I will add, that for aught I have heard or read, the mechanical philosophers have not as yet given a clear and satisfactory account of it.

The second thing is a circular motion upon their own axes, and in some of them also, it is probable, about other points, if we admit the hypothesis of every fixed star being a sun or sunlike body, and having a choir of planets, in like manner, moving about him. These revolutions, we have reason to believe, are as exactly equal and uniform as the earth's are: which could not be were there any place for chance, and did not a providence continually oversee and secure them from all alteration or imminution, which either internal changes in their own parts, or external accidents and occurrences would at one time or other necessarily induce. Without this circular motion of the earth, here could be no living: one hemisphere would be condemned to perpetual cold and darkness, the other continually roasted and parched by the sun-beams. And it is reasonable to think, that this circular motion is as necessary to most other planetary bodies, as it is to the earth. As for the fix'd stars, if they be sun-like bodies, it is probable also each of them moves circularly upon it own axes as the sun doth. But what necessity there is of such a motion, for want of understanding the nature of those bodies, I must confess myself not yet to comprehend; though that it is very great I doubt not, both for themselves, and for the bodies about them.

First, For the celestial or heavenly bodies, the equability and constancy of their motions, the certainty of their periods and revolutions, the conveniency of their order and situations, argue them to be ordained and governed by wisdom and understanding; yea, so much wisdom as man cannot easily fathom or comprehend: for we see by how much the hypothesis of astronomers are more simple and conformable to reason, by so much do they give a better account of the heavenly motions. It is reported of Alphonsus, king of Arragon, I know not whether truly, that when he saw and considered the many eccentrics, epicycles, epicycles upon epicycles, librations, and contrariety of motions, which were requisite in the old hypothesis to give an account of the celestial phenomena, he should presume blasphemously to say, that the universe was a bungling piece; and that if he had been of God's counsel, he could have directed him to have made it better; a speech as rash and ignorant, as daring and profane.

For it was nothing but ignorance of the true process of nature that induced the contrivers of that hypothesis to invent such absurd suppositions, and him to accept them for true, and attribute them to the great Author of the heavenly motions. For in the new hypothesis of the modern astronomers, we see most of those absurdities and irregularities rectified and removed, and I doubt not but they would all vanish, could we certainly discover the true method and process of nature in those revolutions: for seeing in those works of nature which we converse with, we constantly find those axioms true, 'Natura non facit circuitus,' nature doth not fetch a compass when it may proceed in a straight line; and 'Natura nec abundat in superfluis, nec deficit in necessariis,' nature abounds not in what is superfluous, neither is deficient in what is necessary: we may also rationally conclude concerning the heavenly bodies, seeing there is so much exactness observed in the time of their motions, that they punctually come about in the same periods to the hundredth part of a minute, as may beyond exception be demonstrated by comparing their revolutions, surely there is also used the most simple, facile, and convenient way for the performance of them. Among these heavenly bodies,

First, The sun, a vast globe of fire, esteemed by the ancienter and most modest computation above a hundred and sixty times bigger than the earth, the very life of this inferior world, without whose salutary and vivific beams all motion, both animal, vital, and natural, would speedily cease, and nothing be left here below but darkness and death: all plants and animals must needs in a very short time be not only mortified, but together with the surface of land and water, frozen as hard as a flint or adamant: so that of all the creatures of the world, the ancient heathen had most reason to worship him as a God, though no true reason; because he was but a creature, and not God: and we Christians, to think that the service of the animals that live upon the earth, and principally man, was one end of his creation; seeing without him there could no such things have been. This sun, I say, according to the old hypothesis, whirled round about the earth daily with incredible celerity, making night and day by his rising and setting: winter and summer, by his access to the several tropics, creating such a grateful variety of seasons, enlightening all parts of the earth by his beams, and cherishing them by his heat, situate and moved so in respect of this sublunary world (and it is likely also in respect of all the planets about him), that art and counsel could not have designed either to have placed him better, or moved him more conveniently for the service thereof, as I could easily make appear by the inconveniencies that would follow upon the supposition of any other situation and motion, shews forth the great wisdom of him who so disposed and moved him.

Secondly, The moon, a body in all probability somewhat like the earth we live upon, by its constant and regular motion helps us to divide our time, reflects the sun-beams to us, and so by illuminating the air, takes away in some measure the disconsolate darkness of our winter nights; procures, or at least regulates the fluxes and refluxes of the sea, whereby the water is kept in constant motion, and preserved from putrefaction, and so rendered more salutary for the maintenance of its breed, and useful and serviceable for man's conveniences of fishing and navigation; not to mention the great influence it is supposed to have upon all moist bodies, and the growth and increase of vegetables and animals: men generally observing the age of the moon in the planting of all kind of trees, sowing of grain, grafting and inoculating, and pruning of fruit-trees, gathering of fruit, cutting of corn or grass; and thence also making prognostics of weather, because such observations seem to me uncertain. Did this luminary serve to no other ends and uses, as I am persuaded it doth many, especially to maintain the creatures which in all likelihood breed and inhabit there, for which I refer you to the ingenious treatises written by bishop Wilkins and monsieur Fontenelle on that subject, yet these were enough to evince it to be the effect and product of divine wisdom and power.

Thirdly, As for the rest of the planets; besides their particular uses, which are to us unknown, or merely conjectural, their courses and revolutions, their stations and retrogradations, observed constantly so many ages together, in most certain and determinate periods of time, do sufficiently demonstrate that their motions are instituted and governed by counsel, wisdom, and understanding.

Fourthly, The like may be said of the fixed stars, whose motions are regular, equal, and constant. So that we see nothing in the heavens which argues chance, vanity, or error; but on the contrary, rule, order, and constancy; the effects and arguments of wisdom: wherefore as Cicero excellently concludes, 'Cœlestem ergo admirabilem ordinem incredibilemque constantiam, ex qua conservatio et salus omnium omnis oritur, qui vacare mente putat, næ ipse mentis expers habendus est:' 'Wherefore whosoever thinketh that the admirable order and incredible constancy of the heavenly bodies and their motions, whereupon the preservation and welfare of all things doth depend, is not governed by mind and understanding, he himself is to be accounted void thereof. And again, Shall we (saith he), when we see an artificial engine, as a sphere, or dial, or the like, at first sight acknowledge, that it is a work of reason and art?' 'Cum autem impetum cœli admirabili cum celeritate moveri vertique videamus, constantissime conficientem vicissitudines anniversarias, cum summa salute et conservatione rerum omnium, dubitamus, quin ea non solum ratione fiant, sed etiam excellenti quadam divinaque ratione:' 'And can we, when we see the force of the heavens moved and whirled about with admirable celerity, most constantly finishing its anniversary vicissitudes, to the eminent welfare and preservation of all things, doubt at all that these things are performed not only by reason, but by a certain excellent and divine reason.'

To these things I shall add an observation which I must confess myself to have borrowed of the honourable person more than once mentioned already, that even the eclipses of the sun and moon, though they be frightful things to the superstitious vulgar, and of ill influence on mankind, if we may believe the no less superstitious astrologers, yet to knowing men, that can skilfully apply them, they are of great use, and such as common heads could never have imagined: since not only they may on divers occasions help to settle chronology, and rectify the mistakes of historians that writ many ages ago; but which is, though a less wonder, yet of greater utility, they are, as things yet stand, necessary to define with competent certainty, the longitude of places or points on the terraqueous globe, which is a thing of very great moment not only to geography, but to the most useful and important art of navigation. To which may be added, which I shall hereafter mention, that they serve to demonstrate the spherical roundness of the earth.

So that I may well conclude with the Psalmist, Psal. xix. 1. 'The heavens declare the glory of God, and the firmament sheweth his handywork.'

Of terrestrial and inanimate simple bodies.

I come now to consider the terrestrial bodies; I shall say nothing of the whole body of the earth in general, because I reserve that as one of the particulars I shall more carefully and curiously examine.

Terrestrial bodies, according to our method before propounded, are either inanimate or animate, and the inanimate either simple or mixed: simple, as the four elements, fire, water, earth, and air: I call these elements in compliance, as I said before, with the vulgarly-received opinion; not that I think them to be the principles or component ingredients of all other sublunary bodies; I might call them the four great aggregates of bodies of the same species, or four sorts of bodies, of which there are great aggregates. These, notwithstanding they are endued with contrary qualities, and are continually encroaching one upon another, yet they are so balanced, and kept in such an equilibrium, that neither prevaileth over other, but what one gets in one place it loseth in another.

First, Fire cherisheth and reviveth by its heat, without which all things would be torpid and without motion; nay, without fire no life, it being the vital flame residing in the blood that keeps the bodily machine in motion, and renders it a fit organ for the soul to work by. The uses of fire (I do not heres peak of the peripatetics' elementary fire in the concave of the moon, which is but a mere figment, but our ordinary culinary) are in a manner infinite for dressing and preparing of victuals, baked, boiled, and roast; for melting and refining of metals and minerals; for the fusion of glass (a material whose uses are so many, that it is not easy to enumerate them, it serving us to make windows for our houses, drinking-vessels, vessels to contain and preserve all sorts of fermented liquors, distilled waters, spirits, oils, extracts, and other chymical preparations, as also vessels to distil and prepare them in; for looking-glasses, spectacles, microscopes and telescopes, whereby our sight is not only relieved, but wonderfully assisted to make rare discoveries); for making all sorts of instruments for husbandry, mechanic arts, and trade; all sorts of arms or weapons of war, defensive and offensive; for fulminating engines; for burning of lime, baking of bricks, tiles, and all sorts of potter's vessels or earthern ware; for casting and forging metalline vessels and utensils; for distillations, and all chymical operations hinted before in the use of glass; for affording us lights for any work or exercise in winter nights; for digging in mines and dark caverns; and finally by its comfortable warmth securing us from the injuries of cold, or relieving when we have been bitten and benumbed with it. A subject or utensil of so various and inexplicable use, who could have invented and formed, but an infinitely wise and powerful efficient?

Secondly, The air serves us and all animals to breathe in, containing the fuel of that vital flame we speak of, without which it would speedily languish and go out. So necessary is it for us and other land-animals, that without the use of it we could live but very few minutes: nay, fishes and other water-animals cannot abide without the use of it; for if you put fish into a vessel of a narrow mouth full of water, they will live and swim there, not only days and months, but even years. But if with your hand or any other cover you stop the vessel so as wholly to exclude the air, or interrupt its communication with the water, they will suddenly be suffocated; as Rondeletius affirms he often experimented; if you fill not the vessel up to the top, but leave some space empty for the air to take up, and then clap your hand upon the mouth of the vessel, the fishes will presently contend which shall get uppermost in the water, that so they may enjoy the open air; which I have also observed them to do in a pool of water that hath been almost dry in the summer-time, because the air that insinuated itself into the water did not suffice them for respiration. Neither is it less necessary for insects than it is for other animals, but rather more, these having more air-vessels for their bulk by far than they, there being many orifices on each side their bodies for the admission of air, which if you stop with oil or honey, the insect presently dies, and revives no more. This was an observation of the ancients, though the reason of it they did not understand ('Oleo illito insecta omnia exanimantur.' Plin.), which was nothing but the intercluding of the air; for though you put oil upon them, if you put it not upon or obstruct those orifices therewith, whereby they draw the air, they suffer nothing: if you obstruct only some, and not others, the parts which are near and supplied with air from thence, are by and by convulsed, and shortly relaxed and deprived of motion, the rest that were untouched still retaining it. Nay, more than all this, plants themselves have a kind of respiration, being furnished with plenty of vessels for the derivation of air to all their parts, as hath been observed, nay, first discovered, by that great and curious naturalist Malpighius. Another use of the air is to sustain the flight of birds and insects: moreover by its gravity it raises the water in pumps, siphons, and other engines, and performs all those feats which former philosophers through ignorance of the efficient cause attributed to a final, namely, nature's abhorrence of a vacuity or empty space. The elastic or expansive faculty of the air, whereby it dilates itself when compressed (indeed this lower region of it, by reason of the weight of the superincumbent, is always in a compressed state), hath been made use of in the common weather-glasses, in wind-guns, and in several ingenious water-works, and doubtless hath a great interest in many natural effects and operations.

Against what we have said of the necessity of the air for the maintenance of the vital flame, it may be objected, that the fœtus in the womb lives; its heart pulses and its blood circulates; and yet it draws in no air, neither hath the air any access to it. To which I answer, that it doth receive air, so much as is sufficient for it in its present state, from the maternal blood by the *placenta uterina*, or the *cotyledons*. This opinion generally propounded, viz. that the respiration of the dam did serve the fœtus also, or supply sufficient air to it, I have met with in books, but the explicit notion of it I owe to my learned and worthy friend Dr. Edward Hulse, which comparing with mine own anatomical observations, I found so consonant to reason, and highly probable, that I could not but yield a firm assent to it. I say then, that the chief use of the circulation of the blood through the cotyledons of a calf in the womb (which I have often dissected), and by analogy through the placenta uterina in a human fœtus, seems to be the impregnation of the blood with air, for the feeding of the vital flame. For if it were only for nutrition, what need of two such great arteries to convey the blood thither? It would (one might rationally think) be more likely, that as in the abdomen of every animal, so here, there should have been some lacteal veins formed, beginning from the placenta, or cotyledons, which concurring in one common ductus, should at last empty themselves into the *vena cava*. Secondly, I have observed in a calf, the umbilical vessels to terminate in certain bodies divided into a multitude of carneous papillæ, as I may so call them, which are received into so many sockets of the cotyledons growing on the womb; which carneous papillæ may without force or laceration be drawn out of those sockets. Now these papillæ do well resemble the aristæ or radii of a fish's gills, and very probably have the same use to take in the air. So that the maternal blood which flows to the cotyledons, and encircles these papillæ, communicates by them to the blood of the fœtus, the air wherewith itself is impregnate; as the water flowing about the carneous radii of the fish's gills doth the air that is lodged therein to them. Thirdly, that the maternal blood flows most copiously to the placenta uterina in women, is manifest from the great hemorrhagy that succeeds the separation thereof at the birth. Fourthly, after the stomach and intestines are formed, the fœtus seems to take in its whole nourishment by the mouth; there being always found in the stomach of a calf plenty of the liquor contained in the amnios wherein he swims, and fæces in his intestines, and abundance of urine in the allantoides. So that the fœtus in the womb doth live as it were the life of a fish. Lastly, why else should there be such an instant necessity of respiration so soon as ever the fœtus is fallen off from the womb?

I know that if the fœtus be taken out of the womb inclosed in the secundines, it will continue to live, and the blood to circulate for a considerable time, as Dr. Harvey observes. The reason whereof I conceive to be, because the blood still circulates through the cotyledons or placenta, which are now exposed to the open air, and so from thence receives sufficient supplies thereof, to continue its gentle motion, and feed the vital flame. But when upon exclusion of the young the umbilical vessels are broken, and no more air is received that way, the plastic nature, to preserve the life of the animal, speedily raises the lungs, and draws into them air in great abundance, which causes a sudden and mighty accension in the blood; to the maintenance whereof a far greater quantity of air is requisite, than would serve to feed the mild and languid flame before.

This way we may give a facile and very probable account of it, to wit, because receiving no more communications of air from its dam or mother, it must needs have a speedy supply from without, or else extinguish and die for want of it; being not able to live longer without air at its first birth, than it can do afterward.

Upon this occasion give me leave to discourse a little concerning the air's insinuating itself into the water. I say, therefore, that the air, at least that part of it which is the aliment of fire, and fuel of the vital flame in animals, easily penetrates the body of water exposed to it, and diffuseth itself through every part of it. Hence it is that we find fish in subterraneous rivers, and fossil fish in the earth itself; which can no more live without air there than in the open waters: hence the

D

miners, when they come once at water, are out of all danger of damps. You will say, How gets the air into the water in subterraneous rivers, and into the earth to the fossil fishes? I answer, the same way that the water doth: which I suppose to be by its upper superficies; the water descending by pores and passages that there it finds into chinks and veins, and by a confluence of many of them by degrees swelling into a stream, the air accompanies and follows it by a constant succession. As for fossil fishes, some make their way into the earth up the veins of water opening into the banks of rivers, where they lie till they grow so great that they cannot return: in which veins they find air enough to serve their turn, needing not much by reason that they lie still, and move but little. Others in times of floods are left in the meadows, and with the water sink into the earth at some holes and pores that the water finds or makes, by which also they are supplied with air. The reason why the miners are out of danger of damps when they come to water, I conceive is, because then presently the air that stagnated in the shafts sinks into the water, and fresh air descends and succeeds, and so there is a circulation; in the same manner as by the sinking of an air-shaft the air hath liberty to circulate, and carry out the steams both of the miners' breath and the damps, which would otherwise stagnate there. Indeed, though there were no damps, yet the nitrous part of the air being spent and consumed by the breathing of the miners, the remaining part would be rendered altogether unfit for respiration, unless new and fresh air could succeed.

And here methinks appears a necessity of bringing in the agency of some superintendent intelligent being, be it a plastic nature, or what you will. For what else should put the diaphragm, and all the muscles serving to respiration, in motion all of a sudden so soon as ever the fœtus is brought forth? Why could they not have rested as well as they did in the womb? What aileth them that they must needs bestir themselves to get in air to maintain the creature's life? Why could they not patiently suffer it to die? That the air of itself could not rush in, is clear; for that on the contrary there is required some force to remove the incumbent air, and make room for the external to enter. You will say, the spirits do at this time flow to the organs of respiration, the diaphragm and other muscles which concur to that action, and move them. But what rouses the spirits which were quiescent during the continuance of the fœtus in the womb? Here is no appearing impellent but the external air, the body suffering no change but of place, out of its close and warm prison into the open and cool air. But how or why that should have such an influence upon the spirits, as to drive them into those muscles electively, I am not subtle enough to discern. As for the respiration of the chick in the egg, I suppose the air not only to be included in the white, but also to be supplied through the shell and membranes.

Thirdly, Water is one part, and that not the least of our sustenance, and that affords the greatest share of matter in all productions; being not (as it exists in the world) a simple and unmixed body, but containing in it the principles or minute component particles of all bodies. To speak nothing of those inferior uses of washing and bathing, dressing and preparing victuals. But if we shall consider the great concepticula and congregations of water, and the distribution of it all over the dry land in springs and rivers, there will occur abundant arguments of wisdom and understanding. The sea, what infinite variety of fishes doth it nourish! Psal. civ. 25. in the verse next to my text; 'The earth is full of thy riches: so is this great and wide sea, wherein are things creeping innumerable, both small and great beasts,' &c. How doth it exactly compose itself to a level or equal superficies, and with the earth make up one spherical roundness? How doth it constantly observe its ebbs and flows, its spring and neap-tides, and still retain its saltness so convenient for the maintenance of its inhabitants, serving also for the uses of man for navigation, and the convenience of carriage? That it should be defined by shores and strands and limits, I mean at first, when it was natural to it to overflow and stand above the earth. All these particulars declare abundance of wisdom in their primitive constitution. This last the Psalmist takes notice of in the 6, 7, 8, and 9th verses of this psalm. Speaking of the earth at the first creation, he saith, 'Thou coveredst it with the deep as with a garment, the waters stood above the mountains. At thy rebuke they fled, at the voice of thy thunder they hasted away (the mountains ascend, the valleys descend) unto the place thou hast prepared for them. Thou hast set a bound that they may not pass over: that they turn not again to cover the earth.' But what need was there (may some say) that the sea should be made so large, that its superficies should equal, if not exceed that of the dry land? Where is the wisdom of the Creator, in making so much useless sea, and so little dry land, which would have been far more beneficial and serviceable to mankind? Might not at least half the sea have been spared, and added to the land, for the entertainment and maintenance of men, who by their continual striving and fighting to enlarge their bounds, and encroaching upon one another, seem to be straitened for want of room?

To this objection against the wisdom of God in thus dividing sea and land, Mr. Keil in his examination of Dr. Burnet's 'Theory of the Earth,' p. 92, 93. thus answers, 'This, as most other of the atheist's arguments, proceeds from a deep ignorance of natural philosophy. For if there were but half the sea that now is, there would be also but half the quantity of vapours, and consequently we could have but half so many rivers as now there are to supply all the dry land we have at present, and half as much more. For the quantity of vapours which are raised, bears a proportion to the surface whence they are raised, as well as to the heat which raised them. The wise Creator therefore did so prudently order it, that the sea should be large enough to supply vapours sufficient for all the land, which it would not do if it were less than now it is.'

But against this it may be objected, why should not all the vapours which are raised out of the sea fall down again into it by rain? Is there not as much reason that the vapours which are exhaled out of the earth should be carried down to the sea, as that those raised out of the sea be brought up upon the dry land? If some by winds be driven from the sea up land, others by the same cause will be blown down from land to sea, and so balancing one another, they will in sum fall equally upon sea and land; and consequently the sea contribute nothing to the watering of the earth, or the maintenance of rivers.

To which I answer, That as to the watering of the earth there needs no supply from the sea, there being sufficient water exhaled out of itself to do that; there is no more returned upon it by rain so as to rest upon it, than an equivalent quantity to what was raised out of it.

But the rivers must be supplied other ways. Our opinion is, that they have their supply from rain and vapours. The question is, whence these vapours are brought? We answer, from the sea. But what brings them up from the sea? I answer, the winds: and so I am arrived at the main difficulty. Why should not the winds carry them that are exhaled out of the earth down to the sea, as well as bring them up upon the earth which are raised from the sea? Or which is all one, why should not the winds blow indifferently from sea and land? To which I answer, that I must needs acknowledge myself not to comprehend the reason hereof. God is truly said, Psal. cxxxv. 7. 'To bring the wind out of his treasuries.' But the matter of fact is most certain, that the winds do bring abundantly more vapours up from the sea than they carry down thither.

First, Because otherwise there can no account be given of floods. It is clear, that floods with us proceed from rain; and it is often a vast quantity of water they carry down to the sea. Whence come those vapours which supply all this water? I hope those who bring up springs and rivers from the great abyss, will not bring those vapours, which unite into drops and descend in rain from thence too. Should they rise from the dry land only, they would soon render it dry indeed; more parched than the deserts of Lybia. We should quickly come to an end of floods, and of rain too, if nothing were returned from the sea again, not to mention that the sea must needs in such a case overflow its shores, and enlarge its bounds.

But this way there is an easy account to be given. It is clear, that the sun doth exhale vapours both from the sea and land; and that the superficies of sea and land is sufficient to yield vapours for rain, rivers, and floods, when heated to such a degree as the sun heats it. So that there wants only wind to bring up so great a proportion of vapours from the sea as may afford water for the floods: that is, so much as is returned back again to the sea.

Some may perchance demand, to what purpose serve the floods? What use is there of them? I answer, to return back to the sea the surplusage of water after the earth is sated with rain. It may be farther asked, what need more rain be poured upon the earth than is sufficient to water it? I reply, that the rain brings down from the mountains and higher grounds a great quantity of earth, and in times of floods spreads it upon the meadows and levels, rendering them thereby so fruitful as to stand in need of no culture or manuring. So we see the land of Egypt owes its great fertility to the annual overflowing of the river Nilus: and it is likely the countries bordering upon the river of Ganges may receive the like benefit by the overflowing thereof. Moreover, all rain-water contains in it a copious sediment of terrestrial matter, which by standing it precipitates, and is not a simple elementary water. This terrestrial matter serves for the nourishment of plants, and not the water itself, which is but a vehicle to derive this nutriment to all the parts of the plant: and therefore the more rain, the more of this nutritious matter may be precipitated upon the earth, and so the earth rendered more fruitful. Besides all this, it is not unlikely, that the rain-water may be endued with some vegetating or prolific virtue, derived from some saline or oleose particles it contains. For we see that aquatic plants, which grow in the very water, do not thrive and flourish in dry summers, when they are not also watered with the dew of heaven.

Secondly, another argument to prove, that the winds bring up more vapours from the sea than to the sea. This appears from the trees which grow on and near the sea-shores all along the western coast of England, whose heads and boughs I have observed to run out far to Landward, but towards the sea to be so snubbed by the winds, as if their boughs and leaves had been pared off on that side.

It is also observed that the western wind, which is the most violent and boisterous of all with us in England, which comes from off the great Atlantic ocean, is of longest continuance. Julius Cæsar, in his 5th Book of Commentaries de bello Gallico, saith of it, ' Magnam partem omnis temporis in his locis flare consuevit;' it is wont to blow in these quarters a great part of the whole year.' Which observation holds true at this day, the wind lying in that quarter at least three quarters of the year.

Since this motion of the winds is constant, there is doubtless a constant and settled cause of it, which deserves to be inquired into and searched out by the study and endeavours of the most sagacious naturalists. But however the wind be raised, it may more easily blow from sea to land than from land to sea, because the superficies of the sea being even or level, there is nothing to stop its course; but on the land there are not only woods, but mountains to hinder and divert it.

Having myself seen so much of the bottom of the sea round about the coast of England, and a great part of the Low Countries, of Italy and Sicily, I must needs adhere to what I delivered, that where the bottom of the sea is not rocky, but earth, owze or sand, and that is incomparably the greatest part of it, it is by the motion of the waters, so far as the reciprocation of the sea extends to the bottom, brought to a level; and if it should be now unequal, would in time be levelled again. By level I do not mean so as to have no declivity (for the reciprocation preserves that, the flood hindering the constant carrying down of the bottom) but only to have an equal and uniform descent from the shores to the deeps. Now all those relations of urinators belong only to those places where they have dived, which are always rocky: for there is no reason why they should dive where the bottom is level and sandy. That the motion of the water descends to a good depth, I prove from those plants that grow deepest in the sea, because they all generally grow flat in manner of a fan, and not with branches on all sides like trees; which is so contrived by the providence of nature, for that the edges of them do in that posture with most ease cut the water flowing to and fro; and should the flat side be objected to the stream, it would soon be turned edge-wise by the force of it, because in that site it doth least resist the motion of the water: whereas did the branches of these plants grow round, they would be thrown backward and forward every tide. Nay, not only the herbaceous and woody submarine plants, but also the lithophyta* themselves affect this manner of growing, as I have observed in various kinds of corals and pori. Hence I suspect all those relations concerning trees growing at the bottom of the sea, and bringing forth fruit there: and as for the

* Stone-plants.

Maldiva nut, till better information, I adhere to Garcias's opinion, which may be seen in Clusius. Further I do believe, that in the great depths of the sea there grow no plants at all, the bottom being too remote from the external air, which though it may pierce the water so low, yet I doubt whether in quantity sufficient for the vegetation of plants : nay, we are told that in those deep and bottomless seas there are no fish at all; yet not because there are no plants and insects to feed them, for that they can live upon water alone, Rondeletius's experiment about keeping them in a glass doth undeniably prove, but because their spawn would be lost in those seas, the bottom being too cold for it to quicken there ; or rather because being lighter than the water, there, it would not sink to the bottom, but be buoyed up by it, and carried away to the shallows.

Again, the great use and convenience, the beauty and variety of so many springs and fountains, so many brooks and rivers, so many lakes and standing pools of water, and these so scattered and dispersed all the earth over, that no great part of it is destitute of them, without which it must, without a supply otherways, be desolate and void of inhabitants; afford abundant arguments of wisdom and council. That springs should break forth on the sides of mountains most remote from the sea. That there should way be made for rivers through straits and rocks, and subterraneous vaults, so that one would think that nature had cut a way on purpose to derive the water, which else would overflow and drown whole countries. That the water passing through the veins of the earth, should be rendered fresh and portable, which it cannot be by any percolations we can make, but the saline particles will pass through a tenfold filtre. That

IN THE CREATION.

in some places there should spring forth metallic and mineral waters, and hot baths, and these so constant and permanent for many ages; so convenient for divers medicinal intentions and uses, the causes of which things, or the means and methods by which they are performed, have not been as yet certainly discovered; only in general Pliny's remark may be true, 'Tales sunt aquæ, qualis terra per quam fluunt.' Hence they are cold, hot, sweet, stinking, purgative, diuretic, or ferrugineous, saline, petrifying, bituminose, venenose, and of other qualities.

Lastly, the earth, which is the basis and support of all animals and plants, and affords them the hard and solid part of their bodies, yielding us food and sustenance, and partly also clothing; for I do not think that water supplies man and other animals, or even plants themselves, with their nourishment, but serves chiefly for a vehicle to the alimentary particles, to convey and distribute them to the several parts of the body. Water, as it exists in the world, is not a simple unmixed body, but contains the terrestrial component parts of the bodies of animals and plants: simple elementary water nourishes not at all. How variously is the surface of this earth distinguished into hills, and valleys, and plains, and high mountains, affording pleasant prospects? How curiously clothed and adorned with the grateful verdure of herbs and stately trees, either dispersed and scattered singly, or, as it were, assembled in woods and groves, and all these beautified and illustrated with elegant flowers and fruits, 'Quorum omnium incredibilis multitudo insatiabili varietate distinguitur,' as Tully saith. This also shews forth to them that consider it, both the power and wisdom of God: so that we may conclude with Solomon, Prov. iii. 19. 'The

Lord by wisdom hath founded the earth, by understanding hath he established the heavens.'

But now, if we pass from simple to mixed bodies, we shall still find more matter of admiration, and argument of wisdom. Of these we shall first consider those they call imperfectly mixed, or Meteors.

OF METEORS.

As first of all rain, which is nothing else but water by the heat of the sun divided into very small invisible parts, ascending in the air, till encountering the cold, it be by degrees condensed into clouds, and descends in drops; this, though it be exhaled from the salt sea, yet by this natural distillation is rendered fresh and potable, which our artificial distillations have hitherto been hardly able to effect; notwithstanding the eminent use it would be of to navigators, and the rewards promised to those that should resolve that problem of distilling fresh water out of salt. That the clouds should be so carried about by the winds, as to be almost equally dispersed and distributed, no part of the earth wanting convenient showers, unless when it pleaseth God for the punishment of a nation to withhold rain by a special interposition of his providence; or if any land wants rain, they have a supply some other way; as the land of Egypt, though there seldom falls any rain there, yet hath abundant recompense made it by the annual overflowing of the river. This distribution of the clouds and rain is to me (I say) a great argument of providence and divine disposition; for else I do not see but why there might be in some lands continual successive droughts for many years, till they were quite depopulated; in others as lasting rains, till they were overflown and drowned;

and these, if the clouds moved casually, often happening; whereas since the ancientest records of history we do not read or hear of any such droughts or inundations, unless perhaps that of Cyprus, wherein there fell no rain there for thirty-six years, till the island was almost quite deserted, in the reign of Constantine; which doubtless fell not out without the wise disposition of Providence, for great and weighty reasons.

Again, if we consider the manner of the rain's descent, distilling down gradually and by drops, which is most convenient for the watering of the earth; whereas if it should fall down in a continual stream like a river, it would gall the ground, wash away plants by the roots, overthrow houses, and greatly incommode, if not suffocate animals: if, I say, we consider these things and many more that might be added, we might in this respect also cry out with the Apostle, 'O the depth of the riches, both of the wisdom and knowledge of God!'

Secondly, another meteor is the wind; which how many uses it doth serve to, is not easy to enumerate, but many it doth, viz. to ventilate and break the air, and dissipate noisome and contagious vapours, which otherwise stagnating, might occasion many diseases in animals; and therefore it is an observation concerning our native country, 'Anglia ventosa, si non ventosa venenosa:' to transfer the clouds from place to place, for the more commodious watering of the earth: to temper the excesses of the heat, as they find, who in Brazil, New Spain, the neighbouring islands, and other the like countries near the equator, reap the benefit of the breezes: to fill the sails of ships, and carry them on their voyages to remote countries; which of what eminent advantage it is to mankind, for the procuring and continuing of trade and mutual commerce between the most distant nations, the illustrating every corner of the earth, and the perfecting geography and natural history, is apparent to every man. That the monsoons and trade-winds should be so constant and periodical even to the 30th degree of latitude all round the globe, and that they should so seldom transgress or fall short of those bounds, is a subject worthy of the thoughts of the greatest philosophers. To this may be added the driving about of windmills for grinding of corn, making of oil, draining of pools, raising of water, sawing of wood, fulling of cloth, &c. That it should seldom or never be so violent and boisterous, as to overturn houses; yea, whole cities; to tear up trees by the roots, and prostrate woods; to drive the sea over the lower countries; as, were it the effect of chance, or mere natural causes not moderated by a superior power, it would in all likelihood often do. Hurricanes, spouts, and inundations would be more frequent than they are. All these things declare the wisdom and goodness of him who bringeth the wind out of his treasures.

OF INANIMATE MIXED BODIES.

I PROCEED now to such inanimate bodies as are called *perfecte mixta*, perfectly mixed, improperly enough, they being many of them (for aught I know) as simple as those they call elements. These are stones, metals, minerals, and salts.

In stones, which one would think were a neglected genus, what variety? what beauty and elegance? what constancy in their temper and consistency, in their figures and colours? I shall speak of first some notable qualities wherewith some of them are endued; secondly, the remarkable uses they are of to us. The qualities I shall instance in are, first, colour, which in some of them is most lively, sparkling, and beautiful; the carbuncle or rubine shining with red, the sapphire with blue, the emerald with green, the topaz or chrysolyte of the ancients with a yellow or gold colour, the amethyst as it were tinctured with wine, the opal varying its colours like changeable taffata, as it is diversly exposed to the light. Secondly, hardness, wherein some stones exceed all other bodies, and among them the adamant all other stones, being exalted to that degree thereof, that art in vain endeavours to counterfeit it, the factitious stones of chymists in imitation being easily detected by an ordinary lapidist. Thirdly, figure, many of them shoot into regular figures, as crystal and bastard diamonds into hexagonal; others into those that are more elegant and compounded, as those formed in imitation of the shells of testaceous fishes of all sorts, sharks' teeth and vertebres, &c. If these be originally stones, or primary productions of nature in imitation of shells and fishes' bones, and not the shells and bones themselves petrified, as we have sometimes thought. Some have a kind of vegetation and resemblance of plants, as corals, pori, and fungites, which grow upon the rocks like shrubs: to which I might add our ordinary star-stones and trochites, which I look upon as a sort of rock-plants.

Secondly, for the uses; some serve for building, and many sorts of vessels and utensils; for pillars, and statues, and other carved works in relief, for the temples, ornaments of palaces, porticoes, piazzas, conduits, &c. as freestone and marble; some to burn into lime, as chalk and lime-stone; some, with the mixture of beriglia or kelp, to make glass, as that the Venetians call cuogolo, and common flints, which serve also to strike fire; some to cover houses, as slates; some for marking, as morochthus, and the before-mentioned chalk, which is a πολύχρηστον, serving moreover for manuring land, and some medicinal uses; some to make vessels of which will endure the fire, as that found in the country of Chiavenna near Plurs. To these useful stones I might add the warming-stone, digged in Cornwall, which being once well heated at the fire retains its warmth a great while, and hath been found to give ease and relief in several pains and diseases, particularly in that of the internal hæmorrhoids. I might also take notice that some stones are endued with an electrical or attractical virtue. 'My honoured friend Dr. Tancred Robinson, in his manuscript itinerary of Italy, relates the many various figures he observed naturally delineated and drawn on several sorts of stones digged up in the quarries, caverns, and rocks about Florence, and other parts of Italy, not only representing cities, mountains, ruins, clouds, oriental characters, rivers, woods, animals, but also some plants (as ivy, mosses, maiden-hair, ferns, and such vegetables as grow in those places), so exactly designed and impressed upon several kinds of stones, as though some skilful painters or sculptors had been working upon them: the doctor observes also the wonderful diversity of shapes and colours that ores and other fossils shoot into, resembling almost every thing in nature, for which it seems very difficult to him to assign any cause or principle; in the pyrites alone he believes he himself may have seen at home and abroad above a hundred varieties, and yet he confesses he has been but a rude observer of them. In the diaphanous fossils (as ambers, crystals, agates, &c.) preserved in the cabinets of the great duke of

Tuscany, cardinal Chigi, Settali, Moscardi, and other repositories or museums of that curious country, he takes notice of the admirable diversity of bodies included and naturally imprisoned within them, as flies, spiders, frogs, locusts, bees, pismires, gnats, grasshoppers, drops of liquor, hair, leaves, rushes, moss, seeds, and other herbage; which seem to prove them to have been once in a state of fluidity. The bononia stone digged up in the Appenines is remarkable for its shining quality; the amianthus for its incombustibility; the oculus mundi for its motion and change of colour; the lapis nephriticus, calaminaris, ostiocolla, ætites, &c. for their medicinal uses.'

I might spend much time in the discoursing of the most strange and unaccountable nature and powers of the loadstone, a subject which hath exercised the wits and pens of the most acute and ingenious philosophers; and yet the hypothesis which they have invented to give an account of its admirable phenomena seems to me lame and unsatisfactory. What can we say of the subtlety, activity, and penetrancy of its effluvia, which no obstacle can stop or repel, but they will make their way through all sorts of bodies, firm and fluid, dense and rare, heavy and light, pellucid and opaque? Nay, they will pass through a vacuity or empty space, at least devoid of air and any other sensible body. Its attractive power of iron was known to the ancients; its verticity and direction to the poles of the earth is of later invention; which of how infinite advantage it hath been to these two or three last ages, the great improvement of navigation and advancement of trade and commerce by rendering the remotest countries easily accessible, the noble discovery of a vast continent or new world, besides a multitude of unknown kingdoms and islands, the resolving experimentally those ancient problems of the spherical roundness of the earth; of the being of antipodes, of the habitableness of the torrid zone, and the rendering the whole terraqueous globe circumnavigable, do abundantly demonstrate; whereas formerly they were wont to coast it, and creep along the shores, scarce daring to venture out of the ken of land when they did, having no other guide but the cynosura or pole-star, and those near it, and in cloudy weather none at all.

As for metals, they are so many ways useful to mankind, and those uses so well known to all, that it would be lost labour to say any thing of them: without the use of these we could have nothing of culture or civility: no tillage or agriculture; no reaping or mowing; no ploughing or digging; no pruning or lopping; grafting or incision; no mechanical arts or trades; no vessels or utensils of household-stuff; no convenient houses or edifices; no shipping or navigation. What a kind of barbarous and sordid life we must necessarily have lived, the Indians in the northern part of America are a clear demonstration. Only it is remarkable, that those which are of most frequent and necessary use, as iron, brass, and lead, are the most common and plentiful: others that are more rare, may better be spared, yet are they thereby qualified to be made the common measure and standard of the value of all other commodities, and so to serve for coin or money, to which use they have been employed by all civil nations in all ages.

Now of what mighty importance the use of money is to mankind, the learned and ingenious Dr. Cockburn shews us, in the second part of his Essays concerning the Nature of Christian Faith.

p. 88. Whenever, saith he, the use of money began, it was an admirable contrivance for rewarding and encouraging industry, for carrying on trade and commerce certainly, easily, and speedily, for obliging all to employ their various parts and several capacities for the common good, and engaging every one to communicate the benefit of his particular labour, without any prejudice to himself. Covetousness indeed, or an inordinate love of money, is vicious, and the root of much evil, and ought to be remedied; but the use of money is necessary, and attended with manifold advantages. Where money has not yet taken place, where the use of it hath not yet been introduced, arts and sciences are not cultivated, nor any of those exercises plied, which polite men's spirits, and which abate the uneasiness of life. Men there are brutish and savage, none mind any thing but eating and drinking, and the other acts of brutal nature; their thoughts aspire no higher than merely to maintain their life and breath: like the beasts they walk abroad all the day long, and range about from place to place, only to seek their food. Whatever may be supposed to follow if all were acted with great generosity and true charity, yet according to the present temper of mankind it is absolutely necessary that there be some method and means of commutation, as that of money, for rendering all and every one mutually useful and serviceable.

Now gold and silver by their rarity are wonderfully fitted and accommodated for this use of permutation for all sorts of commodities, or making money of: whereas were they as common and easy to come by as straw or stubble, sand or stones, they would be of no more use for bartering and commerce than they.

And here he goes on to shew the wonderful

providence of God, in keeping up the value of gold and silver, notwithstanding the vast quantities which have been digged out of the earth in all ages, and so continuing them a fit material to make money of. For which I refer to the book.

Of these, gold is remarkable for its admirable ductility and ponderosity, wherein it excels all other bodies hitherto known. I shall only add concerning metals, that they do pertinaciously resist all transmutation; and though one would sometimes think they were turned into a different substance, yet do they but as it were lurk under a larva or vizard, and may be reduced again into their natural form and complexions, in despite of all the tortures of Vulcan or corrosive waters. Note, That this was written above thirty years since, when I thought I had reason to distrust whatever had then been reported or written to affirm the transmutation of metals one into another.

I shall omit the consideration of other minerals, and of salts and earths, because I have nothing to say of their uses, but only such as refer to man, which I cannot affirm to have been the sole or primary end of the formation of them. Indeed, to speak in general of these terrestrial inanimate bodies, they having no such organization of parts as the bodies of animals, nor any so intricate variety of texture, but that their production may plausibly be accounted for by an hypothesis of matter divided into minute particles or atoms naturally indivisible, of various but a determinate number of figures, and perhaps also differing in magnitude, and these moved, and continually kept in motion according to certain established laws or rules; we cannot so clearly discover the uses for which they were created, but may probably conclude, that among other ends, they were made for those for which they serve us and other animals; as I shall more fully make out hereafter. It is here to be noted, that according to our hypothesis, the number of the atoms of each several kind, that is, of the same figure and magnitude, is not nearly equal, but there be infinitely more of some species than of others, as of those that compound those vast aggregates of air, water, and earth, more abundantly than of such as make up metals and minerals: the reason whereof may probably be, because those are necessary to the life and being of man and all other animals, and therefore must be always at hand; these only useful to man, and serving rather his conveniences than necessities. The reason why I affirm the minute component particles of bodies to be naturally indivisible by any agent we can employ, even fire itself, which is the only catholic dissolvent, other menstruums being rather instruments than efficients in all solutions, apt by reason of the figure and smallness of their parts to cut and divide other bodies (as wedges cleave wood) when actuated by fire or its heat, which else would have no efficacy at all (as wedges have not, unless driven by a beetle): the reason, I say, I have already given; I shall now instance in a body whose minute parts appear to be indissoluble by the force of fire, and that is common water, which distil, boil, circulate, work upon how you will by fire, you can only dissolve it into vapour, which when the motion ceases easily returns into water again; vapour being nothing else but the minute parts thereof, by heat agitated and separated one from another. For another instance, some of the most learned and experienced chymists do affirm quicksilver to be intransmutable, and therefore call it 'liquor æternus.' And I am of opinion, that the same holds of all simple bodies, that their component particles are indissoluble, by any natural agent.

We may here note the order and method that metals and minerals observe in their growth, how regularly they shoot, ferment, and as it were vegetate and regenerate; salts in their proper and constant figures, as our ingenious countryman Dr. Jordan observes at large in his Discourse of Baths and Mineral Waters.

OF VEGETABLES OR PLANTS.

I have now done with inanimate bodies, both simple and mixed. The animate are,

First, such as are endued only with a vegetative soul, and therefore commonly called Vegetables or Plants; of which if we consider either their stature and shape, or their age and duration, we shall find it wonderful: for why should some plants rise up to a great height, others creep upon the ground, which perhaps may have equal seeds, nay, the lesser plant many times the greater seed? Why should each particular so observe its kind, as constantly to produce the same leaf for consistency, figure, division, and edging; and bring forth the same kind of flower, and fruit, and seed, and that though you translate it into a soil which naturally puts forth no such kind of plant, so that it is some Λόγος σπερματικὸς,* which doth effect this, or rather some intelligent plastic nature; as we have before intimated: for what account can be given of the determination of the growth and magnitude of plants from mechanical principles, of matter moved without the presidency and guidance of some superior agent? Why may trees not grow

* Seminal form or virtue.

up as high as the clouds or vapours ascend, or if you say the cold of the superior air checks them, why may they not spread and extend their lateral branches so far till their distance from the centre of gravity depress them to the earth, be the tree never so high? How comes it to pass that though by culture and manure they may be highly improved and augmented to a double, treble, nay some a much greater proportion in magnitude of all their parts; yet is this advance restrained within certain limits? There is a *maximum quod sic* which they cannot exceed. You can by no culture or art extend a fennel stalk to the stature and bigness of an oak: then why should some be long lived, others only annual or biennial? How can we imagine that any laws of motion can determine the situation of the leaves, to come forth by pairs or alternately, or circling the stalk, the flowers to grow singly, or in company and tufts, to come forth the bosoms of the leaves and branches, or on the tops of branches and stalks; the figure of the leaves, that they should be divided into so many jags or escallops, and curiously indented round the edges; as also of the flower-leaves, their number and site, the figure and number of the stamina and their apices, the figure of the stile and seed vessel, and the number of cells into which it is divided? That all this be done, and all these parts duly proportioned one to another, there seems to be necessary some intelligent plastic nature, which may understand and regulate the whole economy of the plant: for this cannot be the vegetative soul, because that is material and divisible together with the body: which appears in that a branch cut off of a plant will take root, and grow, and become a perfect plant itself, as we have already observed. I had almost forgotten the complication of the seed-leaves of some plants in the seed, which is so strange, that one cannot believe it to be done by matter, however moved by any laws or rules imaginable. Some of them being so close plaited, and straitly folded up and thrust together within the membranes of the seed, that it would puzzle a man to imitate it, and yet none of the folds sticking or growing together; so that they may easily be taken out of their cases, and spread and extended even with one's fingers.

Secondly, if we consider each particular part of a plant, we shall find it not without it end or use: the roots for its stability and drawing nourishment from the earth; the fibres to contain and convey the sap. Besides which, there is a large sort of vessels to contain the proper and specific juice of the plant: and others to carry air for such a kind of respiration as it needeth; of which we have already spoken. The outer and inner bark in trees serve to defend the trunk and boughs from the excesses of heat and cold and drought, and to convey the sap for the annual augmentation of the tree. For in truth every tree may in some sense be said to be an annual plant, both leaf, flower, and fruit proceeding from the coat that was superinduced over the wood the last year, which coat also never beareth any more, but together with the old wood serves as a form or block to sustain the succeeding annual coat. The leaves before the gemma or bud be explicated to embrace and defend, the flower and fruit, which is even then perfectly formed; afterward to preserve the branches, flowers, and fruit from the injuries of the summer sun, which would too much parch and dry them, if they lay open and exposed to its beams without any shelter; the leaves, I say, qualify and contemper the heat, and serve also to hinder the too hasty evaporation of the moisture about the root; but the principal use of the leaves (as we learn of Seignior Malpighii, Monsieur Perault, and Monsieur Mariotte) is to concoct and prepare the sap for the nourishment of the fruit, and the whole plant, not only that which ascends from the root, but what they take in from without, from the dew, moist air, and rain. This they prove because many trees, if despoiled of their leaves, will die; as it happens sometimes in mulberry-trees, when they are plucked off to feed silkworms. And because if in summer-time you denude a vine-branch of its leaves, the grapes will never come to maturity: but though you expose the grapes to the sun-beams, if you pluck not off the leaves, they will ripen notwithstanding. That there is a regress of the sap in plants from above downwards; and that this descendent juice is that which principally nourisheth both fruit and plant, is clearly proved by the experiments of Seignior Malpighii, and those rare ones of an ingenious countryman of our own,* Thomas Brotherton, esquire, of which I shall mention only one, that is, if you cut off a ring of bark from the trunk of any tree, that part of the tree above the barked ring shall grow and encrease in bigness, but not that beneath.

But whether there be such a constant circulation of the sap in plants as there is of the blood in animals, as they would from hence infer, there is some reason to doubt. I might add hereto the pleasant and delectable, cooling and refreshing shade they afford in the summer-time; which was very much esteemed by the inhabitants of hot countries, who always took great delight and pleasure to sit in the open air, under shady trees; hence that expression so often repeated in scrip-

* Philosop. Transact. Num. 187.

ture, of every man's sitting under his own fig-tree, where also they used to eat; as appears by Abraham's entertaining the angels under a tree, and standing by them when they did eat, Gen. xviii. 8. Moreover the leaves of plants are very beautiful and ornamental. That there is great pulchritude and comeliness of proportion in the leaves, flowers, and fruits of plants, is attested by the general verdict of mankind, as Dr. More and others well observe. The adorning and beautifying of temples and buildings in all ages, is an evident and undeniable testimony of this: for what is more ordinary with architects than the taking in leaves and flowers and fruitage for the garnishing of their work; as the Roman, the leaves of *Acanthus sat.* and the Jewish of palm-trees and pomegranates: and these more frequently than any of the five regular solids, as being more comely and pleasant to behold. If any man shall object, that comeliness of proportion and beauty is but a mere conceit, and that all things are alike handsome to some men who have as good eyes as others; and that this appears by the variation of fashions, which doth so alter men's fancies, that what ere-while seemed very handsome and comely, when it is once worn out of fashion appears very absurd, uncouth, and ridiculous. To this I answer, that custom and use doth much in those things where little of proportion and symmetry shew themselves, or which are alike comely and beautiful, to disparage the one, and commend the other. But there are degrees of things; for (that I may use* Dr. More's words) I dare appeal to any man that is not sunk into so forlorn a pitch of degeneracy, that he is as stupid to these things as the basest beasts, whether, for example, a rightly-cut *Te-*

* Antidote against Atheism, l. 2. §. 5.

traëdrum, *Cube*, or *Icosaëdrum*, have no more pulchritude in them than any rude broken stone, lying in the field or high-ways; or to name other solid figures, which though they be not regular, properly so called, yet have a settled idea and nature, as a cone, sphere, or cylinder, whether the sight of those do not more gratify the minds of men, and pretend to more elegancy of shape than those rude cuttings or chippings of free-stone that fall from the mason's hands, and serve for nothing but to fill up the middle of the wall, as fit to be hid from the eyes of men for their ugliness. And therefore it is observable, that if nature shape any thing but near to this geometrical accuracy, that we take notice of it with much content and pleasure, and greedily gather and treasure it up. As if it be but exactly round, as those spherical stones found in Cuba, and some also in our own land, or have but its sides parallel, as those rhomboideal selenites found near St. Ives in Huntingdonshire, and many other places in England. Whereas ordinary stones of rude and uncertain figures we pass by, and take no notice of at all. But though the figures of these bodies be pleasing and agreeable to our minds, yet (as we have already observed) those of the leaves, flowers, and fruits of trees, more. And it is remarkable, that in the circumscription and complication of many leaves, flowers, fruits, and seeds, nature affects a regular figure. Of a pentagonal or quincunial disposition Sir Thomas Brown of Norwich produces several examples in his discourse about the *quincunx*. And doubtless instances might be given in other regular figures, were men but observant.

The flowers serve to cherish and defend the first and tender rudiments of the fruit: I might also add the masculine or prolific seed contained in the chives or apices of the stamina. These beside the elegancy of their figures, are many of them endued with splendid and lovely colours, and likewise most grateful and fragrant odours. Indeed such is the beauty and lustre of some flowers, that our Saviour saith of the lilies of the field (which some, not without reason, suppose to have been tulips) that Solomon in all his glory was not arrayed like one of these. And it is observed by* Spigelius, that the art of the most skilful painter cannot so mingle and temper his colours, as exactly to imitate or counterfeit the native ones of the flowers of vegetables.

As for the seeds of plants,† Dr. More esteems it an evident sign of divine Providence, that every kind hath its seed : for it being no necessary result of the motion of the matter (as the whole contrivance of the plant indeed is not), and it being of so great consequence, that they have seed for the continuance and propagation of their own species, and also for the gratifying man's art, industry, and necessities (for much of husbandry and gardening lies in this), it cannot but be an act of counsel to furnish the several kinds of plants with their seeds.

Now the seed being so necessary for the maintenance and increase of the several species, it is worthy the observation, what care is taken to secure and preserve it, being in some doubly and trebly defended. As for instance, in the walnut, almond, and plums of all sorts, we have first a thick pulpy covering, then a hard shell, within which is the seed enclosed in a double membrane. In the nutmeg another tegument is added besides all these, *viz.* the mace between the green pericarpium and the hard shell, immediately enclosing the kernel. Neither yet doth the exterior

* Isag. 3d rem Herbariam. † Antid. against Atheism, l. 2. c. 6.

pulp of the fruit or pericarpium serve only for the defence and security of the seed, whilst it hangs upon the plant; but after it is mature and fallen upon the earth, for the stercoration of the soil, and promotion of the growth, though not the first germination of the seminal plant. Hence (as[*] Petrus de Crescentiis tells us) husbandmen to make their vines bear, manure them with vine-leaves, or the husks of exprest grapes, and that they observe those to be the most fruitful, which are so manured with their own: which observation holds true also in all other trees and herbs. But besides this use of the pulp or pericarpium, for the guard and benefit of the seed, it serves also by a secondary intention of nature in many fruits for the food and sustenance of man and other animals.

Another thing worthy the noting in seeds, and argumentative of providence and design, is that pappose plumage growing upon the tops of some of them, whereby they are capable of being wafted with the wind, and by that means scattered and disseminated far and wide.

Furthermore, most seeds having in them a seminal plant perfectly formed, as the young is in the womb of animals, the elegant complication thereof in some species is a very pleasant and admirable spectacle; so that no man that hath a soul in him can imagine or believe it was so formed and folded up without wisdom and providence. But of this I have spoken already.

Lastly, The immense smallness of some seeds, not to be seen by the naked eye, so that the number of seeds produced at once in some one plant; as for example, reedmace ('tipha palustris') harts-tongue, and many sorts of ferns, may amount to a million, is a convincing argument

[*] Agric. l. 2. c. 6.

of the infinite understanding and art of the former of them.

And it is remarkable that such mosses as grow upon walls, the roofs of houses and other high places, have seeds so excessively small, that when shaken out of their vessels they appear like vapour of smoke, so that they may either ascend of themselves, or by an easy impulse of the wind be raised up to the tops of houses, walls, or rocks. And we need not wonder how the mosses got thither, or imagine they sprung up spontaneously there.

I might also take notice of many other particulars concerning vegetables, as First, That because they are designed for the food of animals, therefore nature hath taken more extraordinary care, and made more abundant provision for their propagation and increase; so that they are multiplied and propagated not only by the seed, but many also by the root, producing offsets or creeping under ground, many by strings and wires running above ground, as strawberry and the like, some by slips or cuttings, and some by several of these ways. And for the security of such species as are produced only by seed, it hath endued all seed with a lasting vitality, that so if by reason of excessive cold, or drought, or any other accident, it happen not to germinate the first year, it will continue its fecundity, I do not say two or three, nor six or seven, but even twenty or thirty years; and when the impediment is removed, the earth in fit case, and the season proper, spring up, bear fruit, and continue its species. Hence it is that plants are sometimes lost for a while in places where they formerly abounded; and again, after some years, appear new: lost either because the springs were not proper for their germination, or because the land was fallowed, or because plenty of weeds or other herbs prevented their coming up, and the like; and appearing again when these impediments are removed. Secondly, That some sorts of plants, as vines, all sorts of pulse, hops, briony, all pomiferous herbs, pumpions, melons, gourds, cucumbers, and divers other species, that are weak and unable to raise or support themselves, are either endued with a faculty of twining about others that are near, or else furnished with claspers and tendrils, whereby, as it were with hands, they catch hold of them, and so ramping upon trees, shrubs, hedges, or poles, they mount up to a great height, and secure themselves and their fruit. Thirdly, That others are armed with prickles and thorns, to secure them from the browsing of beasts, as also to shelter others that grow under them. Moreover they are hereby rendered very useful to man, as if designed by nature to make both quick and dead hedges and fences. The great naturalist Pliny hath given an ingenious account of the providence and design of nature in thus arming and fencing them in these words: 'Inde,' speaking of nature, ' excogitavit aliquas aspectu hispidas, tactu truces, ut tantum non vocem ipsius naturæ fingentis illas rationemque reddentis exaudire videamur, ne se depascat avida quadrupes, ne procaces manus rapiant, ne neglecta vestigia obterant, ne insidens ales infringat; his muniendo aculeis telisque armando, remediis ut salva ac tuta sint. Ita hoc quoque quod in iis odimus hominum causa excogitatum est.'

It is worthy the noting, that wheat, which is the best sort of grain, of which the purest, most savoury, and wholesome bread is made, is patient of both extremes, heat and cold, growing and bringing its seed to maturity, not only in temperate countries, but also on one hand in the cold and northern, viz. Scotland, Denmark, &c. on the other, in the hottest and most southerly, as Egypt, Barbary, Mauritania, the East Indies, Guinea, Madagascar, &c. scarce refusing any climate.

Nor is it less observable, and not to be commemorated without acknowledgment of the divine benignity to us, that (as Pliny rightly notes) nothing is more fruitful than wheat, 'Quod ei natura,' saith he, ('rectius naturæ parens) tribuit, quod eo maxime hominem alit, utpote cum e modio, si et aptum solum, quale in Byzacio Africæ campo, centeni quinquaginta modii reddentur. Misit ex eo loco Divo Augusto procurator ejus ex uno grano (vix credibile dictu) 400 paucis minus germina: misit et Neroni similiter 360 stipulas ex uno grano.' 'Which fertility nature (he should have said, the author of nature) hath conferred upon it, because it feeds man chiefly with it. One bushel, if sown in a fit and proper soil, such as is Byzacium, a field of Africa, yielding 150 of annual increase. Augustus's procurator sent him from that place 400 within a few blades, springing from the same grain: and to Nero were sent thence 360.' If Pliny, a heathen, could make this fertility of wheat argumentative of the bounty of God to man, making such plentiful provision for him of that which is of most pleasant taste and wholesome nourishment, surely it ought not to be passed over by us Christians without notice taking and thanksgiving.

As for the signatures of plants, or the notes impressed upon them as indices of their virtues, though some lay great stress upon them,* accounting them strong arguments to prove that

* D. More Antid. l. 2. c. 6.

some understanding principle is the highest original of the works of nature; as indeed they were, could it certainly be made appear that there were such marks designedly set upon them; because all that I find mentioned and collected by authors, seem to me to be rather fancied by men, than designed by nature to signify or point out any such virtues or qualities as they would make us believe. I have elsewhere, I think upon good grounds, rejected them; and finding no reason as yet to alter my opinion, I shall not farther insist on them. Howbeit I will not deny but that the noxious and malignant plants do, many of them, discover something of their nature by the sad and melancholic visage of their leaves, flowers, and fruit. And that I may not leave that head wholly untouched, one observation I shall add relating to the virtues of plants, in which I think there is something of truth, that is, that there are, by the wise disposition of Providence, such species of plants produced in every country as are most proper and convenient for the meat and medicine of the men and animals that are bred and inhabit there. Insomuch that Solenander writes, that from the frequency of the plants that spring up naturally in any region he could easily gather what endemial diseases the inhabitants thereof were subject to: so in Denmark, Friezland, and Holland, where the scurvy usually reigns, the proper remedy thereof, scurvy-grass, doth plentifully grow.

OF BODIES ENDUED WITH A SENSITIVE SOUL, OR ANIMALS.

I PROCEED now to the consideration of animate bodies endued with a sensitive soul, called animals. Of these I shall only make some general observations, not curiously consider the parts of each particular species, save only as they serve for instances or examples.

First of all, because it is the great design of Providence to maintain and continue every species, I shall take notice of the great care and abundant provision that is made for the securing this end. 'Quanta ad eam rem vis, ut in suo quæque genere permaneat?' Cic. Why can we imagine all creatures should be made male and female but to this purpose? why should there be implanted in each sex such a vehement and inexpugnable appetite of copulation? Why in viviparous animals, in the time of gestation should the nourishment be carried to the embryon in the womb, which at other times goeth not that way? When the young is brought forth, how comes all the nourishment then to be transferred from the womb to the breasts or paps, leaving its former channel, the dam at such time being for the most part lean and ill-favoured? To all this I might add, as a great proof and instance of the care that is taken, and provision made for the preservation and continuance of the species, the lasting fecundity of the animal seed or egg in the females of man, beasts, and birds. I say the animal seed, because it is to me highly probable, that the females, as well of beasts as birds, have in them from their first formation the seeds of all the young they will afterward bring forth, which when they are all spent and exhausted, by what means soever, the animal becomes barren and effete. These seeds in some species of animals continue fruitful, and apt to take life by the admixture of the male seed fifty years or more, and in some birds fourscore or an hundred. Here I cannot omit one very remarkable observation I find in Cicero: 'Atque ut intelligamus (saith he) nihil horum esse fortuitum, sed hæc omnia providæ solertisque naturæ, quæ multiplices fœtus procreant, ut sues, ut canes, his mammarum data est multitudo ; quas easdem paucas habent eæ bestiæ, quæ pauca gignunt.' 'That we may understand that none of these things (he had been speaking of) is fortuitous but that all are the effects of provident and sagacious nature; multiparous quadrupeds, as dogs, as swine, are furnished with a multitude of paps : whereas those beasts which bring forth few, have but a few.'

That flying creatures of the greater sort, that is, birds, should all lay eggs, and none bring forth live young, is a manifest argument of divine providence, designing thereby their preservation and security, that their might be the more plenty of them ; and that neither the birds of prey, the serpent, nor the fowler, should straiten their generations too much. For if they had been viviparous, the burthen of their womb, if they had brought forth any competent number at a time, had been so great and heavy, that their wings would have failed them, and they become an easy prey to their enemies: or if they had brought but one or two at a time, they would have been troubled all the year long with feeding their young, or bearing them in their womb. Dr. More, Antid. Atheism. l. 2. c. 9.

This mention of feeding their young puts me in mind of two or three considerable observations referring thereto.

First, Seeing it would be for many reasons inconvenient for birds to give suck, and yet no less inconvenient, if not destructive, to the chicken upon exclusion all of a sudden to make so great a change in its diet, as to pass from liquid to hard food, before the stomach be gradually consolidated, and by use strengthened and habituated to grind and concoct it, and its tender and pappy flesh fitted to be nourished by such strong and solid diet; and before the bird be by little and little accustomed to use its bill, and gather it up, which at first it doth but very slowly and imperfectly; therefore nature hath provided a large yolk in every egg, a great part whereof remaineth after the chicken is hatched, and is taken up and enclosed in its belly, and by a channel made on purpose received by degrees into the guts, and serves instead of milk to nourish the chick for a considerable time; which nevertheless mean while feeds itself by the mouth a little at a time, and gradually more and more, as it gets a perfecter ability and habit of gathering up its meat, and its stomach is strengthened to macerate and concoct it, and its flesh hardened and fitted to be nourished by it.

Secondly, That birds which feed their young in the nest, though in all likelihood they have no ability of counting the number of them, should yet (though they bring but one morsel of meat at a time, and have not fewer, it may be, than seven or eight young in the nest together, which at the return of their dams, do all at once with equal greediness, hold up their heads and gape) not omit or forget one of them, but feed them all; which, unless they did carefully observe and retain in memory which they had fed, which not, were impossible to be done: this I say, seems to me most strange and admirable, and beyond the possibility of a mere machine to perform.

Another experiment I shall add to prove, that though birds have not an exact power of numbering, yet have they of distinguishing many from few, and knowing when they come near to a certain number: and that is, that when they have laid such a number of eggs, as they can conveniently cover and hatch, they give over and begin to sit; not because they are necessarily determined to such a number; for that they are not, as is clear, because they have an ability to go on and lay more at their pleasure. Hens, for example, if you let their eggs alone, when they have laid fourteen or fifteen, will give over and begin to sit, whereas if you daily withdraw their eggs they will go on to lay five times that number. [Yet some of them are so cunning, that if you leave but one egg, they will not lay to it, but forsake their nest.] This holds not only in domestic and mansuete birds, for then it might be thought the effect of cicuration or institution, but also in the wild; for my honoured friend Dr. Martin Lister informed me, that of his own knowledge one and the same swallow, by the subtracting daily of her eggs, proceeded to lay nineteen successively, and then gave over: as I have* elsewhere noted. Now that I am upon this subject of the number of eggs, give me leave to add a remarkable observation referring thereto, *viz.* that birds and such oviparous creatures as are long-lived, have eggs enough at first conceived in them to serve them for many years laying, probably for as many as they are to live, allowing such a proportion for every year, as will serve for one or two incubations; whereas insects, which are to breed but once, lay all their eggs at once, have they never so many. Now had these things been governed by chance, I see no reason why it should constantly fall out so.

Thirdly, the marvellous speedy growth of birds that are hatched in nests, and fed by the old ones there, till they be fledged and come almost to their full bigness; at which perfection they arrive within the short term of about one fortnight, seems to me an argument of providence designing

* Preface to Mr. Willughby's Ornithol.

thereby their preservation, that they might not lie long in a condition exposed to the ravine of any vermin that may find them, being utterly unable to escape or shift for themselves.

Another and no less effectual argument may be taken from the care and providence used for the hatching and rearing their young. And first, they search out a secret and quiet place, where they may be secure and undisturbed in their incubation; then they make themselves nests, every one after his kind, that so their eggs and young may lie soft and warm, and their exclusion and growth be promoted. These nests some of them so elegant and artificial, that it is hard for man to imitate them and make the like. I have seen nests of an Indian bird so artificially composed of the fibres, I think, of some roots, so curiously interwoven and platted together, as is admirable to behold: which nests they hang on the end of the twigs of trees over the water, to secure their eggs and young from the ravage of apes and monkeys, and other beasts that might else prey upon them. After they have laid their eggs, how diligently and patiently do they sit upon them till they be hatched, scarce affording themselves time to go off to get them meat? nay, with such an ardent and impetuous desire of sitting are they inspired, that if you take away all their eggs, they will sit upon an empty nest: and yet one would think that sitting were none of the most pleasant works. After their young are hatched, for some time they do almost constantly brood them under their wings, lest the cold and sometimes perhaps the heat should harm them. All this while also they labour hard to get them food, sparing it out of their own bellies, and pining themselves almost to death rather than they should want. Moreover it is admirable to observe, with what courage they are at that time inspired, that they will even venture their own lives in defence of them. The most timorous, as hens and geese, become then so courageous, as to dare to fly in the face of a man that shall molest or disquiet their young, which would never do so much in their own defence. These things being contrary to any motions of sense, or instinct of self-preservation, and so eminent pieces of self-denial, must needs be the works of Providence, for the continuation of the species and upholding of the world: especially if we consider that all this pains is bestowed upon a thing which takes no notice of it, will render them no thanks for it, nor make them any requital or amends; and also, that after the young is come to some growth, and able to shift for itself, the old one retains no such στοργὴ to it, takes no further care of it, but will fall upon it, and beat it indifferently with others. To these I shall add three observations more relating to this head. The first borrowed of Dr. Cudworth, 'System,' p. 69. One thing necessary to the conservation of the species of animals; that is, the keeping up constantly in the world a due numerical proportion between the sexes of male and female, doth necessarily infer a superintending Providence. For did this depend only upon mechanism, it cannot well be conceived, but that in some ages or other there should happen to be all males, or all females; and so the species fail. Nay, it cannot well be thought otherwise, but that there is in this a providence, superior to that of the plastic or spermatick nature, which hath not so much of knowledge and discretion allowed to it, as whereby to be able to govern this affair.

The second of Mr. Boyle, in his treatise of the 'High veneration Man's Intellect owes to

God,' p. 32. that is, the conveniency of the season (or time of year) of the production of animals, when there is proper food and entertainment ready for them. 'So we see, that according to the usual course of nature, lambs, kids, and many other living creatures, are brought into the world at the spring of the year ; when tender grass, and other nutritive plants are provided for their food. And the like may be observed in the production of silkworms (yea, all other eruca's, and many insects more) whose eggs, according to nature's institution, are hatched when mulberry-trees begin to bud, and put forth those leaves, whereon those precious insects are to feed. The aliments being tender, whilst the worms themselves are so, and growing more strong and substantial, as the insects increase in vigour and bulk.' To these I shall add another instance, that is, of the wasp, whose breeding is deferred till after the summer solstice, few of them appearing before July : whereas one would be apt to think the vigorous and quickening heat of the sun in the youth of the year should provoke them to generate much sooner. (Provoke them, I say, because every wasps'-nest is begun by one great mother-wasp, which overlives the winter, lying hid in some hollow-tree or other 'Latibulum.') Because then, and not till then, pears, plums, and other fruit, designed principally for their food, begin to ripen.

The third is mine own, that all insects which do not themselves feed their young, nor treasure up provision in store for their sustenance, lay their eggs in such places as are most convenient for their exclusion, and where, when hatched, their proper food is ready for them. So, for example, we see two sorts of white butterflies fastening their eggs to cabbage-leaves, because they are a fit aliment for the caterpillars that come of them: whereas, should they affix them to the leaves of a plant improper for their food, such caterpillars must needs be lost, they choosing rather to die than to taste of such plants. For that kind of insect (I mean caterpillars) hath a nice and delicate palate, some of them feeding only upon one particular species of plant, others on divers indeed, but those of the same nature and quality; utterly refusing them of a contrary. Like instances might be produced in the other tribes of insects; it being perpetual in all, if not hindered or imprisoned, electively to lay their eggs in places where they are seldom lost or miscarry, and where they have a supply of nourishment for their young so soon as they are hatched, and need it. Whereas should they scatter them carelessly and indifferently in any place, the greatest part of the young would in all likelihood perish soon after their exclusion for want of food, and so their numbers continually decreasing, the whole species in a few years in danger to be lost: whereas no such thing, I dare say, hath happened since the first creation.

It is here very remarkable, that those insects, for whose young nature hath not made provision of sufficient sustenance, do themselves gather and lay up in store for them. So for example, the bee, the proper food of whose 'eulæ'* is honey, or perchance ' erithace' (which we English bee-bread), neither of which viands is any where to be found amassed by nature in quantities sufficient for their maintenance, doth herself with unwearied diligence and industry, flying from flower to flower, collect and treasure them up.

To these I shall now add an observation of Mr. Lewenhoeck's, concerning the sudden

* Bee-maggot.

growth of some sorts of insects, and the reason of it.

It is, saith he, a wonderful thing, and worthy the observation, in flesh-flies, that a fly-maggot, in five days' space after it is hatched, arrives at its full growth and perfect magnitude. For if to the perfecting of it there were required, suppose a month's time or more (as in some other maggots is needful), it is impossible that about the summer-solstice any such flies should be produced; because the fly-maggots have no ability to search out any other food than that wherein they are placed by their dams. Now this food, suppose it be flesh, fish, or the entrails of beasts, lying in the fields, exposed to the hot sun-beams, can last but a few days in case and condition to be a fit aliment for these creatures, but will soon be quite parched and dried up. And therefore the most wise Creator hath given such a nature and temperament to them, that within a very few days they attain to their just growth and magnitude. Whereas on the contrary, other maggots, who are in no such danger of being straitened for food, continue a whole month or more before they give over to eat, and cease to grow. He proceeds farther to tell us, that some of these fly-maggots which he fed daily with fresh meat, he brought to perfection in four days' time; so that he conceives that in the heat of summer the eggs of a fly, or the maggots contained in them, may in less than a month's space run through all their changes, and come to perfect flies, which may themselves lay eggs again.

Secondly, I shall take notice of the various strange instincts of animals; which will necessarily demonstrate, that they are directed to ends unknown to them, by a wise superintendant. As, 1. That all creatures should know how to defend themselves, and offend their enemies; where their natural weapons are situate, and how to make use of them. A calf will so manage his head as though he would push with his horns, even before they shoot. A boar knows the use of his tushes: a dog of his teeth; a horse of his hoofs; a cock of his spurs: a bee of her sting; a ram will butt with his head, yea, though he be brought up tame, and never saw that manner of fighting. Now, why another animal which hath no horns should not make a shew of pushing, or no spurs, of striking with his legs, and the like, I know not, but that every kind is providentially directed to the use of its proper and natural weapons. 2. That those animals that are weak, and have neither weapons nor courage to fight, are for the most part created swift of foot or wing, and so being naturally timorous, are both willing and able to save themselves by flight. 3. That poultry, partridge, and other birds, should at the first sight know birds of prey, and make sign of it by a peculiar note of their voice to their young, who presently thereupon hide themselves. That the lamb should acknowledge the wolf its enemy, though it had never seen one before, as is taken for granted by most naturalists, and may, for aught I know, be true, argues the providence of nature, or more truly the God of nature, who for their preservation hath put an instinct into them. 4. That young animals, so soon as they are brought forth, should know their food. As for example: such as are nourished with milk, presently find their way to the paps, and suck at them, whereas none of those that are not designed for that nourishment ever offer to suck, or to seek out any such food. Again, 5. That such creatures as are whole-footed, or fin-toed, viz. some birds and quadrupeds, are naturally directed to go into the water and swim there, as we see ducklings, though hatched and led by a hen, if she brings them to the brink of a river or pond of water, they presently leave her, and in they go, though they never saw any such thing done before; and though the hen clucks and calls, and doth what she can to keep them out: this Pliny takes notice of, Hist. Nat. lib. 10. cap. 55. in these words, speaking of hens: 'Super omnia est anatum ovis subditis atque exclusis admiratio, primo non plane agnoscentis fœtum; mox incertos incubitus solicite convocantis; postremo lamenta circa piscinæ stagna, mergentibus se pullis natura duce.' So that we see every part in animals is fitted to its use, and the knowledge of this use put into them. For neither do any sort of web-footed fowls live constantly upon the land, or fear to enter the water, nor any land-fowl so much as attempt to swim there. 6. Birds of the same kind make their nests of the same materials, laid in the same order, and exactly of the same figure, so that by the sight of the nest one may certainly know what bird it belongs to. And this they do, though living in distant countries, and though they never saw, nor could see any nest made, that is, though taken out of the nest and brought up by hand; neither were any of the same kind ever observed to make a different nest, either for matter or fashion. This, together with the curious and artificial contexture of such nests, and their fitness and convenience for the reception, hatching, and cherishing the eggs and young of their respective builders (which we have before taken notice of), is a great argument of a superior author of their and others' natures, who hath endued them with these instincts, whereby they are, as it were, acted and driven to bring about ends which themselves aim not at (so far as we can discern), but are directed to; for (as Aristotle observes) οὔτε τέχνη, οὔτε ζητήσαντα, οὔτε βουλευσάμενα, ποιεῖ, 'They act not by any art, neither do they inquire, neither do they deliberate about what they do.' And, therefore, as Dr. Cudworth saith well, they are not masters of that wisdom according to which they act, but only passive to the instincts and impresses thereof upon them. And indeed to affirm, that brute animals do all these things by a knowledge of their own, and which themselves are masters of, and that without deliberation and consultation, were to make them to be endued with a most perfect intellect, far transcending that of human reason: whereas it is plain enough, that brutes are not above consultation, but below it; and that these instincts of nature in them, are nothing but a kind of fate upon them.

The migration of birds from a hotter to a colder country, or a colder to a hotter, according to the seasons of the year, as their nature is, I know not how to give an account of, it is so strange and admirable. What moves them to shift their quarters? You will say, The disagreeableness of the temper of the air to the constitution of their bodies, or want of food. But how come they to be directed to the same place yearly, though sometimes but a little island, as the Soland Goose to the Basse of Edinburgh Frith, which they could not possibly see, and so it could have no influence upon them that way? The cold or the heat might possibly drive them in a right line from either, but that they should impel land birds to venture over a wide ocean, of which they can see no end, is strange and unaccountable: one would think that the sight of so much water, and present fear of drowning, should overcome the sense of hunger, or disagreeableness of the temper of the air. Besides, how come they to steer their course aright to their several quarters, which before the compass was invented was hard for man himself to do, they being not able, as I noted before, to see them at a distance? Think we that the quails, for instance, could see quite across the Mediterranean sea? And yet, it is clear, that they fly out of Italy into Africa: lighting many times on ships in the midst of the sea, to rest themselves when tired and spent with flying. That they should thus shift places, is very convenient for them, and accordingly we see they do it; which seems to be impossible they should, unless themselves were endued with reason, or directed and acted by a superior intelligent cause.

The like may be said of the migration of divers sorts of fishes. As for example; the salmon, which from the sea yearly ascends up a river sometimes four or five hundred miles, only to cast their spawn, and secure it in banks of sand, for the preservation of it till the young be hatched or excluded, and then return to sea again. How these creatures, when they have been wandering a long time in the wide ocean, should again find out and repair to the mouths of the same rivers, seems to me very strange, and hardly accountable, without recourse to instinct, and the direction of a superior cause. That birds, seeing they have no teeth for the mastication and preparation of their food, should for the more convenient comminution of it in their stomachs or gizzards, swallow down little pebble-stones, or other hard bodies, and because all are not fit or proper for that use, should first try them in their bills, to feel whether they be rough or angular, for their turns; which if they find them not to be, they reject them. When these, by the working of the stomach, are worn smooth, or too small for their use, they avoid them by siege, and pick up others. That these are of great use to them for the grinding of their meat, there is no doubt; and I have observed in birds, that have been kept up in houses, where they could get no pebbles, the very yolks of their eggs have changed colour, and become a great deal paler, than theirs who have had their liberty to go abroad.

Besides, I have observed in many birds, the gullet, before its entrance into the gizzard, to be much dilated, and thick set, or, as it were, granulated, with a multitude of glandules, each whereof was provided with its excretory vessel, out of which, by an easy pressure, you might squeeze a juice or pap, which served for the same use which the saliva doth in quadrupeds; that is, for the maceration and dissolution of the meat into a chyle. For that the saliva, notwithstanding its insipidness, hath a notable virtue of macerating and dissolving bodies, appears by the effects it hath in killing of quicksilver, fermenting of dough like leaven or yeast, taking away warts, and curing other cutaneous distempers; sometimes exulcerating the jaws, and rotting the teeth.

Give me leave to add one particular more concerning birds, which some may perchance think too homely and indecent to be mentioned in such a discourse as this; yet, because it is not below the providence of nature, and designed for cleanliness, and some great men have thought it worth the observing, I need not be ashamed to take notice of it; that is, that in young birds that are fed in the nest, the excrement that is voided at one time is so viscid, that it hangs together in a great lump, as if it were inclosed in a film, so that it may easily be taken up, and carried away by the old bird in her bill. Besides, by a strange instinct, the young bird elevates her hinder parts so high, for the most part, that she seldom fails to cast what comes from her clear over the side of the nest. So we see here is a double provision made to keep the nest clean, which if it were defiled with ordure, the young ones must necessarily be utterly marred and ruined. 7. The bee, a creature of the lowest form of animals, so that no man can suspect it to have any considerable measure of understanding, or to have knowledge of, much less to aim at any end, yet makes her combs and cells with that geometrical accuracy, that she must needs be acted by an instinct implanted in her by the wise Author of nature. For, first, she plants them in a perpendicular posture, and so close together, as with conveniency they may, beginning at the top, and working downwards, that so no room may be lost in the hive, and that she may have easy access to all the combs and cells. Besides, the combs being wrought double, that is, with cells on each side, a common bottom or partition-wall, could not in any other site have so conveniently, if at all, received or contained the honey. Then she makes the particular cells most geometrically and artificially, as the famous mathematician Pappus demonstrates in the preface to his third book of Mathematical Collections. First of all (saith he, speaking of the cells), it is convenient that they be of such figures as may cohere one to another, and have common sides, else there would be empty spaces left between them to no use, but to the weakening and spoiling of the work, if any thing should get in there. And therefore though a round figure be most capacious for the honey, and most convenient for the bee to creep into, yet did she not make choice of that, because then there must have been triangular spaces left void. Now there are only three rectilineous and ordinate figures which can serve to this purpose; and inordinate, or unlike ones, must have been not only less elegant and beautiful, but unequal. [Ordinate figures are such as have all their sides, and all their angles equal.] The three ordinate figures are, triangles, squares, and hexagons: for the space about any point may be filled up either by six equilateral triangles, or four squares, or three hexagons; whereas three pentagons are too little, and three heptagons too much. Of these three the bee makes use of the hexagon, both because it is more capacious than either of the other, provided they be of equal compass, and so equal matter spent in the construction of each: and, secondly, Because it is most commodious for the bee to creep into: and, lastly, Because in the other figures more angles and sides must have met together at the same point, and so the work could not have been so firm and strong. Moreover, the combs being double, the cells on each side the partition are so ordered, that the angles on one side insist upon the centres of the bottom of the cells on the other side, and not angle upon, or against angle; which also must needs contribute to the strength and firmness of the work. These cells she fills with honey for her winter provision, and curiously closes them up with covers of wax, that keep the included liquor from spilling, and from external injuries; as Mr. Boyle truly observes, 'Treatise of Final Causes,' p. 169. Another sort of bee I have observed it may be called the tree-bee, whose industry is admirable in making provision for her young. First, She digs round vaults or burrows, 'cuniculos,' in a rotten or decayed tree, of a great length, in them she builds or forms her cylindrical nests or cases, resembling cartridges, or a very narrow thimble, only in proportion longer, of pieces of rose or other leaves, which she shares off with her mouth, and plaits and joins close together by some glutinous substance. These cases she fills with a red pap, of a thinner consistence than an electuary, of no pleasant taste, which where she gathers I know not: and, which is most remarkable, she forms these cases, and stores them with this provision, before she hath any young one hatched, or so much as an egg laid; for on the top of the pap she lays one egg, and then closes up the vessel with a cover of leaves. The enclosed egg soon becomes an eula, or maggot, which feeding upon the pap till it comes to its full growth, changes to a nympha, and after comes out a bee. Another insect, noted for her seeming prudence, in making provision for the winter, proposed by Solomon to the sluggard for his imitation, is the ant, which (as all naturalists agree) hoards up grains of corn against the winter for her sustenance: and is reported by some* to bite off the germen of them, lest they should sprout by the moisture of the earth, which I look upon as a mere fiction; neither should I be forward to credit the former relation, were it not for the authority of the Scripture, because I could never observe any such storing up of grain by our country ants.

Yet is there a quadruped taken notice of even by the vulgar, for laying up in store provision for the winter; that is, the squirrel, whose hoards of nuts are frequently found, and pillaged by them.

The beaver is by credible persons, eye-witnesses, affirmed to build him houses for shelter

* Plin. l. ii. c. 30.

and security in winter-time: see Mr. Boyle, of 'Final Causes.' p. 173.

Besides these I have mentioned, a hundred others may be found in books relating especially to physic; as that dogs, when they are sick, should vomit themselves by eating grass: that swine should refuse meat so soon as they feel themselves ill, and so recover by abstinence: that the bird ibis should teach men the way of administering clysters, Plin. lib. 8. cap. 27. The wild goats of Dictamnus for drawing out of darts, and healing wounds. the swallow the use of celandine for repairing the fight, &c. ibid. Of the truth of which, because I am not fully satisfied, I shall make no inference from them.

Thirdly, I shall remark the care that is taken for the preservation of the weak, and such as are exposed to injuries; and preventing the increase of such as are noisome and hurtful: for as it is a demonstration of the divine power and magnificence to create such a variety of animals, not only great but small, not only strong and courageous, but also weak and timorous; so is it no less argument of his wisdom to give to these means, and the power and skill of using them, to preserve themselves from the violence and injuries of those. That of the weak some should dig vaults and holes in the earth, as rabbits, to secure themselves and their young; others should be armed with hard shells; others with prickles, the rest that have no such armature, should be endued with great swiftness or pernicity: and not only so, but some also have their eyes stand so prominent, as the hare, that they can see as well behind as before them, that so they may have their enemy always in their eye; and long, hollow, moveable ears, to receive and convey the least sound, or that which comes from far, that they be not suddenly surprised or taken (as they say) napping. Moreover, it is remarkable, that in this animal, and in the rabbit, the muscles of the loins and hind legs are extraordinarily large in proportion to the rest of the body, or those of other animals, as if made on purpose for swiftness, that they may be able to escape the teeth of so many enemies as continually pursue and chase them. Add hereto the length of their hind legs, which is no small advantage to them, as is noted by dame Julian Barns, in an ancient dialogue in verse between the huntsman and his man: the man there asks his master, what is the reason, why the hare when she is near spent makes up a hill? The master answers, that nature has made the hinder legs of the hare longer than the fore legs; by which means she climbs the hill with much more ease than the dogs, whose legs are of equal length, and so leaves the dogs behind her, and many times escapes away clear, and saves her life. This last observation, I must confess myself to have borrowed out of the papers of my honoured friend Mr. John Aubrey, which he was pleased to give me a sight of.

I might here add much concerning the wiles and ruses, which these timid creatures make use of to save themselves, and escape their persecutors, but that I am somewhat diffident of the truth of those stories and relations: I shall only aver what myself have sometimes observed of a duck, when closely pursued by a water-dog; she not only dives to save herself (which yet she never does but when driven to an exigent, and just ready to be caught, because it is painful and difficult to her), but when she comes up again, drings not her whole body above water, but only her bill, and part of her head, holding the rest underneath, that so the dog, who the mean time

IN THE CREATION. 121

turns round and looks about him, may not espy her, till she have recovered breath.

As for sheep, which have no natural weapons or means to defend or secure themselves, neither heels to run, nor claws to dig; they are delivered into the hand, and committed to the care and tuition of man, and serving him for divers uses, are nourished and protected by him; and so enjoying their beings for a time, by this means propagate and continue their species: so that there are none destitute of some means to preserve themselves, and their kind; and these means so effectual, that notwithstanding all the endeavours and contrivances of man and beast to destroy them, there is not to this day one species lost of such as are mentioned in histories, and consequently and undoubtedly neither of such as were at first created.

Then for birds of prey, and rapacious animals, it is remarkable what Aristotle observes, that they are all solitary, and go not in flocks, Γαμψωνύχων οὐδὲν ἀγελαῖον. No birds of prey are gregarious. Again, that such creatures do not greatly multiply, τῶν γαμψωνύχων ὀλιγοτόκα πάντα. They for the most part breeding and bringing forth but one or two, or at least a few young ones at once: whereas they that are feeble and timorous are generally multiparous; or if they bring forth but few at once, as pigeons, they compensate that by their often breeding, viz. every month but two throughout the year; by this means providing for the continuation of their kind. But for the security of these rapacious birds, it is worthy the noting, that because a prey is not always ready, but perhaps they may fail of one some days; nature has made them patient of a long inedia, and besides, when they light upon one, they gorge themselves so therewith, as to suffice for their nourishment for a considerable time.

Fourthly, I shall note the exact fitness of the parts of the bodies of animals to every one's nature and manner of living. A notable instance of which we have in the swine, a creature well known, and therefore what I shall observe of it is obvious to every man. His proper and natural food being chiefly the roots of plants, he is provided with a long and strong snout; long, that he might thrust it to a sufficient depth into the ground, without offence to his eyes; strong and conveniently formed for the rooting and turning up the ground. And besides, he is endued with a notable sagacity of scent, for the finding out such roots as are fit for his food: hence in Italy, the usual method for the finding and gathering of trufles, or subterraneous mushrooms (called by the Italians *tartusali,* and in Latin *tubera terræ*), is by tying a cord to the hind leg of a pig, and driving him before them into such pastures as usually produce that kind of mushroom, and observing where he stops and begins to root, and there digging, they are sure to find a trufle; which when they have taken up, they drive away the pig to search for more. So I have myself observed, that in pastures where there are earthnuts to be found up and down in several patches, though the roots lie deep in the ground, and the stalks be dead long before and quite gone, the swine will by their scent easily find them out, and root only in those places where they grow.

This rooting of the hog in the earth calls to mind another instance of like nature, that is the porpesse, which as his English name porpesse, i. e. *porc pesce,*[*] imports, resembles the hog, both in the strength of his snout, and also in the man-

[*] Swine-fish.

ner of getting his food by rooting. For we found the stomach of one we dissected, full of sand-eels, or launces, which for the most part lie deep in the sand, and cannot be gotten but by rooting or digging there. We have seen the country-people in Cornwall, when the tide was out, to fetch them out of the sand with iron hooks thrust down under them, made for that purpose.

Furthermore, that very action for which the swine is abominated, and looked upon as an unclean and impure creature, namely wallowing in the mire, is designed by nature for a very good end and use, viz. not only to cool his body, for the fair water would have done that as well, nay better, for commonly the mud and mire in summer-time is warm; but also to destroy lice, fleas, and other noisome and importunate insects that are troublesome and noxious to him. For the same reason do all the poultry-kind, and divers other birds, bask themselves in the dust in summer-time and hot weather, as is obvious to every one to observe.

2. A second and no less remarkable instance I shall produce, out of Dr. More's 'Antidote against Atheism,' lib. 2. cap. 10. in a poor and contemptible quadruped, the mole.

First of all (saith he), her dwelling being under ground, where nothing is to be seen, nature hath so obscurely fitted her with eyes, that naturalists can scarcely agree whether she hath any sight at all or no. [In our observation, moles have perfect eyes, and holes for them through the skin, so that they are outwardly to be seen by any one that shall diligently search for them; though indeed they are exceedingly small, not much bigger than a great pin's-head.] But for amends, what she is capable of for her defence and warning of danger, she has very eminently conferred upon her; for she is very quick of hearing [doubtless her subterraneous vaults are like trunks to convey any sound a great way]. And then her short tail, and short legs, but broad fore-feet armed with sharp claws, we see by the event to what purpose they are, she so swiftly working herself under ground, and making her way so fast in the earth, as they that behold it cannot but admire it. Her legs therefore are short, that she need dig no more than will serve the mere thickness of her body : and her fore feet are broad, that she may scoop away much earth at a time : and she has little or no tail, because she courses it not on the ground like a rat or mouse, but lives under the earth, and is fain to dig herself a dwelling there; and she making her way through so thick an element, which will not easily yield as the water and air do, it had been dangerous to draw so long a train behind her; for her enemy might fall upon her rear, and fetch her out before she had perfected and got full possession of her works : which being so, what more palpable argument of Providence than she?

Another instance in quadrupeds might be the tamandua, or ant-bear, described by Marcgrave and Piso, who saith of them, that they are nightwalkers, and seek their food by night. Being kept tame they are fed with flesh, but it must be minced small, because they have not only a slender and sharp head and snout, but also a narrow and too thless mouth; their tongue is like a great lutestring (as big as a goose quill), round, and in the greater kind (for there are two species more than two feet long, and therefore lies doubled in a channel between the lower parts of the cheeks. This, when hungry, they trust forth, being well moistened, and lay upon the trunks of trees, and when it is covered with ants, suddenly draw it back into their mouths; if the ants lie so deep that they cannot come at them, they dig up the earth with their long and strong claws, wherewith for that purpose their fore feet are armed. So we see how their parts are fitted for this kind of diet, and no other; for the catching of it, and for the eating of it, it requiring no comminution by the teeth; as appears also in the chameleon, which is another quadruped that imitates the tamandua in this property of darting out the tongue to a great length, with wonderful celerity, and for the same purpose too of catching of insects.

Besides these quadrupeds, there are a whole genus of birds, called pici martii, or woodpeckers, that in like manner have a tongue which they can shoot forth to a very great length, ending in a sharp stiff bony tip, dented on each side; and at pleasure thrust it deep into the holes, clefts, and crannies of trees, to stab and draw out coffi, or any other insects lurking there, as also into ant-hills, to strike and fetch out the ants and their eggs. Moreover, they have short, but very strong legs, and their toes stand two forwards, two backwards, which disposition (as Aldrovandus well notes) nature, or rather the wisdom of the Creator, hath granted to woodpeckers, because it is very convenient for the climbing of trees, to which also conduces the stiffness of the feathers of their tails, and their bending downward, whereby they are fitted to serve as a prop for them to lean upon, and bear up their bodies. As for the chameleon, he imitates the woodspite, not only in the make, motion, and use of his tongue for striking ants, flies, and other insects; but also in the site of his toes, whereby he is wonderfully qualified to run upon trees, which he doth with that swiftness, that one would think he flew, whereas upon the ground he walks very clumsily and ridiculously. A full description of the outward and inward parts of this animal, may be seen at the end of Panarolus's Observat. It is to be noted, that the chameleon, though he hath teeth, uses them not for chewing his prey, but swallows it immediately.

I shall add two instances more in birds, and those are,

1. The swallow; whose proper food is small beetles, and other insects flying about in the air; as we have found by dissecting the stomachs both of old ones and nestlings: which is wonderfully fitted for the catching of these animalcules; for she hath long wings, and a forked tail, and small feet, whereby she is, as it were, made for swift flight, and enabled to continue long upon the wing, and to turn nimbly in the air: and she hath also an extraordinary wide mouth, so that it is very hard for any insect that comes in her way to escape her. It is thought to be a sign of rain, when this bird flies low near to the ground; in which there may be some truth; because the insects which she hunts may at such times, when the superior air is charged with vapours, have a sense of it, and descend near the earth. Hence, when there are no more insects in the air, as in winter-time, those birds do either abscond, or betake themselves into hot countries.

2. The colymbi, or douchers, or loons, whose bodies are admirably fitted and conformed for diving under water: being covered with a very thick plumage; and the superficies of their feathers so smooth and slippery, that the water cannot penetrate or moisten them: whereby their bodies are defended from the cold, the water being kept at a distance; and so poised, that by a light impulse they may easily ascend

in it. Then their feet are situate in the hindmost part of their body, whereby they are enabled, shooting their feet backwards, and striking the water upwards, to plunge themselves down into it with great facility, and likewise to move forwards therein. Then their legs are made flat and broad, and their feet cloven into toes with appendant membranes on each side; by which configuration they easily cut the water, and are drawn forward, and so take their stroke backwards; and besides, I conceive, that by means of this figure their feet being moved to the right and left-hand, serve them as a rudder to enable them to turn under water : for some conceive that they swim easier under water than they do above it. How they raise themselves up again; whether their bodies emerge of themselves by their lightness, or whether by striking against the bottom, in manner of a leap, or by some peculiar motion of their legs, I cannot determine : that they dive to the bottom is clear, for that in the stomachs both of the greater and lesser kinds we found grass and other weeds; and in the lesser kind nothing else; though both prey upon fish. Their bills also are made straight and sharp for the easier cutting of the water, and striking their prey. Could we see the motions of their legs and feet in the water, then we should better comprehend how they ascend, descend, and move to and fro ; and discern how wisely and artificially their members are formed and adapted to those uses.

2. In birds all the members are most exactly fitted for the use of flying. First, the muscles which serve to move the wings are the greatest and strongest, because much force is required to the agitation of them ; the underside of them is also made concave, and the upper convex, that

they may be easily lifted up, and more strongly beat the air, which by this means doth more resist the descent of their body downward. Then the trunk of their body doth somewhat resemble the hull of a ship; the head, the prow, which is for the most part small, that it may the more easily cut the air, and make way for their bodies; the train serves to steer, govern, and direct their flight, and however it may be held erect in their standing or walking, yet it is directed to lie almost in the same plane with their backs, or rather a little inclining, when they fly. That the train serves to steer and direct their flight, and turn their bodies like the rudder of a ship, is evident in the kite, who by a light turning of his train, moves his body which way he pleases. 'Iidem videntur artem gubernandi docuisse caudæ flexibus, in cœlo monstrante natura quod opus esset in profundo,' Plin. lib. 10. cap. 10. 'They seem to have taught men the art of steering a ship by the flexures of their tails; nature shewing in the air what was needful to be done in the deep.' And it is notable that Aristotle truly observes, that whole-footed birds, and those that have long legs, have for the most part short tails; and therefore whilst they fly, do not, as others, draw them up to their bellies, but stretch them at length backwards, that they may serve to steer and guide them instead of tails. Neither doth the tail serve only to direct and govern the flight, but also partly to support the body, and keep it even; wherefore, when spread, it lies parallel to the horizon, and stands not perpendicular to it, as fishes do. Hence birds that have no tails, as some sorts of colymbi, or douckers, fly very inconveniently with their bodies almost erect.

To this I shall add farther, that the bodies of

birds are small in comparison of quadrupeds, that they may more easily be supported in the air during their flight; which is a great argument of wisdom and design: else why should not we see species of pegasi, or flying horses, of griffins, of harpies, and a hundred more, which might make a shift to live well enough, notwithstanding they could make no use of their wings. Besides, their bodies are not only small, but of a broad figure, that the air may more resist their descents, they are also hollow and light; nay, their very bones are light: for though those of the legs and wings are solid and firm, yet have they ample cavities, by which means they become more rigid and stiff; it being demonstrable, that a hollow body is more stiff and flexible than a solid one of equal substance or matter. Then the feathers also are very light, yet their shafts hard and stiff, as being either empty or filled with a light and spongy substance; and their webs are not made of continued membranes, for then had a rupture by any accident been made in them, it couldnot have been consolidated, but of two series of numerous plumulæ, or contiguous filaments, furnished all along with hooks on each side, whereby catching hold on one another, they stick fast together; so that when they are ruffled or discomposed, the bird with her bill can easily preen them, and reduce them to their due position again. And for their firmer cohesion, the wise and bountiful Author of nature hath provided and placed on the rump two glandules, having their excretory vessels, round which grow feathers in form of a pencil, to which the bird turning her head, catches hold upon them with her bill, and a little compressing the glandules, squeezes out and brings away therewith an oily pap or liniment, most fit and proper for the inunction of the feathers, and causing their little filaments more strongly to cohere. And is not this strange and admirable, and argumentative of Providence, that there should be such an unguent or pap prepared, such an open vessel to excern it into, to receive and retain it; that the bird should know where it is situate, and how, and to what purposes to use it? And because the bird is to live many years, and the feathers in time would, and must necessarily be worn and shattered, nature hath made provision for the casting and renewing of them yearly. Moreover, those large bladders or membranes, extending to the bottoms of the bellies of birds, into which the breath is received, conduce much to the alleviating of the body, and facilitating the flight: for the air received into these bladders, is by the heat of the body extended into twice or thrice the dimensions of the external, and so must needs add a lightness to the body. And the bird, when she would descend, may either compress this air by the muscles of the abdomen, or expire as much of it as may enable her to descend swifter, or slower, as she pleases. I might add the use of the feathers in cherishing and keeping of the body warm; which, the creature being of small bulk, must needs stand it in great stead against the rigour of the cold. And for this reason we see, that water-fowls, which were to swim and sit long upon the cold water, have their feathers very thick set upon their breasts and bellies, and besides a plentiful down there growing, to fence against the cold of the water, and to keep off its immediate contact.

That the tails of all birds in general do not conduce to their turning to the right and left, according to the common opinion, but rather for their ascent and descent, some modern philosophers have observed and proved by experiment, for that if you pluck off, for instance, a pigeon's tail, she will nevertheless with equal facility turn to and fro : which upon second thoughts, and farther consideration, I grant to be true, in birds whose tails are pointed, and end in a right line : but in those that have forked tails, autopsy convinceth us, that it has this use; and therefore they pronounce too boldly of all in general. For it is manifest to sight, that the fork-tailed kite by turning her train side-ways, elevating one horn, and depressing the other, turns her whole body. And doubtless the tail hath the same use in swallows, who make the most sudden turns in the air of any birds, and have all of them forked tails.

3. As for fishes, their bodies are long and slender, or else thin for the most part, for their more easy swimming and dividing the water. The wind-bladder, wherewith most of them are furnished, serves to poise their bodies, and keep them equiponderant to the water; which else would sink to the bottom, and lie grovelling there, as hath, by breaking the bladder, been experimentally found. By the contraction and dilatation of this bladder, they are able to raise or sink themselves at pleasure, and continue in what depth of water they list. The fins made of gristly spokes or rays, connected by membranes, so that they may be contracted or extended like women's fans, and furnished with muscles for motion, serve partly for progression, but chiefly to hold the body upright; which appears in that when they are cut off, it waves to and fro, and so soon as the fish dies, the belly turns upward. The great strength by which fishes dart themselves forward with incredible celerity, like an arrow out of a bow, lies in their tails, their fins

mean time, lest they should retard their motion, being held close to their bodies. And therefore almost the whole musculous flesh of the body is bestowed upon the tail and back, and serves for the vibration of the tail, the heaviness and corpulency of the water requiring a great force to divide it.

In cetaceous fishes, or, as the Latins call them, sea-beasts,* the tail hath a different position from what it hath in all other fishes, for whereas in these it is erected perpendicular to the horizon, in them it lies parallel thereto, partly to supply the use of the hinder-pair of fins which these creatures lack, and partly to raise and depress the body at pleasure. For it being necessary that these fishes should frequently ascend to the top of the water to breathe, or take in, and let out the air: it was fitting and convenient, that they should be provided with an organ to facilitate their ascent and descent as they had occasion. And as for their turning of their bodies in the water, they must perform that as birds do, by the motion of one of their fins, while the other is quiescent. It is no less remarkable in them, that their whole body is encompassed round with a copious fat, which our fishermen call the blubber, of a great thickness; which serves partly to poise their bodies, and render them equiponderant to the water; partly to keep off the water at some distance from the blood, the immediate contact whereof would be apt to chill it; and partly also for the same use that clothes serve us, to keep the fish warm by reflecting the hot steams of the body, and so redoubling the heat, as we have before noted; for we see by experience, that fat bodies are nothing near so sensible of the impressions of cold as lean. And

* Belluæ marinæ.

I have observed fat hogs to have lain abroad in the open air, upon the cold ground in winter-nights, whereas the lean ones have been glad to creep into their cots, and lie upon heaps to keep themselves warm.

I might here take notice of those amphibious creatures, which we may call aquatic quadrupeds (though one of there is that hath but two feet, viz. the manati, or sea-cow) the beaver, the otter, the phoca, or sea-calf, the water-rat, and the frog, the toes of whose feet are joined by membranes, as in water-fowls, for swimming; and who have very small ears, and ear-holes, as the cetaceous fishes have, for hearing in the water.

To this head belongs the adapting of the parts that minister to generation in the sexes one to another; and in creatures that nourish their young with milk, the nipples of the breast to the mouth and organs of suction; which he must needs be wilfully blind and void of sense, that either discerns not, or denies to be intended and made one for the other. That the nipples should be made spongy, and with such perforations, as to admit passage to the milk when drawn, otherwise to retain it; and the teeth of the young either not sprung, or so soft and tender as not to hurt the nipples of the dam, are effects and arguments of providence and design.

A more full description of the breasts and nipples I meet with, in a book of that ingenious anatomist and physician, Antoninus Nuck, intitled 'Adenographia Curiosa,' cap. 2. He makes the breasts to be nothing but glandules of that sort they call 'conglomeratæ,' made up of an infinite number of little knots or kernels, each whereof hath its excretory vessel, or lactiferous duct; three or four, or five of these presently meet, and join into one small trunk; in like manner do the adjacent glandules meet and unite; several of these lesser trunks or branches concurring, make up an excretory vessel of a notable bigness, like to that of the pancreas, but not so long, yet sufficiently large to receive and retain a good quantity of milk; which, before it enters the nipple is again contracted, and straightened to that degree, that it will scarce admit a small bristle. Who now can be so impudent as to deny, that all this was contrived and designed purposely to retain the milk, that it should not flow out of itself, but easily be drawn out by pressure and suction; or to affirm that this fell out accidentally, than which there could not have been a more ingenious contrivance for the use to which it is employed, invented by the wit of man.

To this head of the fitness of the parts of the body to the creature's nature and manner of living, belongs that observation of Aristotle, τῶν ὀρνίθων ὅσα μὲν γαμψώνυχα σαρκοφάγα πάντα. Such birds as have crooked beaks and talons, are all carnivorous; and so of quadrupeds, καρχαρόδοντα, *carnivora omnia*. All that have serrate teeth, are carnivorous. This observation holds true concerning all European birds, but I know not but that parrots may be an exception to it. Yet it is remarkable, that such birds as are carnivorous have no gizzard, or musculous, but a membraneous stomach, that kind of food needing no such grinding or comminution as seeds do, but being torn into strings, or small flakes, by the beak, may be easily concocted by a membraneous stomach.

To the fitness of all the parts and members of animals to their respective uses, may also be referred another observation of the same Aristotle, Πάντα τὰ ζῶα ἀρτίους ἔχει πόδας. All animals have even feet, not more on one side than another; which if they had, would either hinder their walking, or hang by not only useless, but also burthensome. For though a creature might make a limping shift to hop, suppose with three feet, yet nothing so conveniently or steadily to walk, or run, or indeed to stand. So that we see, Nature hath made choice of what is most fit, proper, and useful. They have also not only an even number of feet, answering by pairs one to another, which is as well decent as convenient; but those too of an equal length, I mean the several pairs; whereas were those on one side longer than they on the other, it would have caused an inconvenient halting or limping in their going.

I shall mention but one more observation of Aristotle, that is, Πτηνὸν μόνον οὐδὲν, there is no creature only volatile, or no flying animal but hath feet as well as wings, a power of walking or creeping upon the earth; because there is no food, or at least not sufficient food for them to be had always in the air; or if in hot countries we may suppose there is, the air being never without store of insects flying about in it, yet could such birds take no rest, for having no feet, they could not perch upon trees, and if they should alight upon the ground, they could by no means raise themselves any more, as we see those birds which have but short feet, as the swift and martinet with difficulty do. Besides, they would want means of breeding, having nowhere to lay their eggs, to sit, hatch, or brood their young. As for the story of the manucodiata, or bird of Paradise, which in the former age was generally received and accepted for true, even by the learned, it is now discovered to be a fable, and rejected and exploded by all men: those birds being well known to have legs and feet as well as others, and those not short, small, nor feeble ones, but sufficiently great and strong, and armed with crooked talons, as being the members of birds of prey.

It is also very remarkable, that all flying insects should be covered with shelly scales, like armour, partly to secure them from external violence, from injuries by blows and pressures: partly to defend their tender muscles from the heat of the sun-beams, which would be apt to parch and dry them up, being of small bulk; partly also to restrain the spirits, and to prevent their evaporation.

I shall now add another instance of the wisdom of nature, or rather the God of nature, in adapting the parts of the same animal one to another, and that is the proportioning the length of the neck to that of the legs. For seeing terrestrial animals, as well birds as quadrupeds, are endued with legs, upon which they stand, and wherewith they transfer themselves from place to place, to gather their food, and for other conveniences of life, and so the trunk of their body must needs be elevated above the superficies of the earth, so that they could not conveniently either gather their food or drink, if they wanted a neck, therefore Nature hath not only furnished them therewith, but with such an one as is commensurable to their legs, except here the elephant, which hath indeed a short neck; for the excessive weight of his head and teeth, which to a long neck would have been unsupportable, but is provided with a trunk, wherewith, as with a hand, he takes up his food and drink, and brings it to his mouth. I say, the necks of birds and quadrupeds are commensurate to their legs, so that they which have long legs have long necks, and they that have short legs short ones, as is seen in the crocodile, and all lizards; and those that have no legs, as they do not want necks, so neither have they any, as fishes. This equality between the length of the legs and neck, is especially seen in beasts that feed constantly upon grass, whose necks and legs are always very near equal; very near I say, because the neck must necessarily have some advantage, in that it cannot hang perpendicularly down, but must incline a little. Moreover, because this sort of creatures must needs hold their heads down in an inclining posture for a considerable time together, which would be very laborious and painful for the muscles; therefore on each side the ridge of the vertebres of the neck, Nature hath placed an ἀπονεύρωσις, or nervous ligament of a great thickness and strength, apt to stretch and shrink again as need requires, and void of sense, extending from the head (to which, and the next vertebres of the neck it is fastened at that end) to the middle vertebres of the back (to which it is knit at the other) to assist them to support the head in that posture, which aponeurosis is taken notice of by the vulgar by the name of fixfax, or pack-wax, or whit-leather. It is also very observable in fowls that wade in the water, which having long legs, have also necks answerably long. Only in these too there is an exception, exceeding worthy to be noted, for some water-fowl, which are palmipeds, or whole-footed, have very long necks, and yet but short legs, as swans and geese, and some Indian birds; wherein we may observe the admirable providence of Nature. For such birds as were to search and gather their food, whether herbs or insects, in the bottom of pools and deep waters, have long necks for that purpose, though their legs, as is most convenient for swimming, be but short. Whereas there are no land-fowl to be seen with short legs, and long necks, but all have their necks in length commensurate to their legs. This instance is the more considerable, because the atheists' usual flam will not here help them out. For (say they) there were many animals of disproportionate parts, and of absurd and uncouth shapes, produced at first, in the infancy of the world; but because they could not gather their food to perform other functions necessary to maintain life, they soon perished, and were lost again. For these birds we see, can gather their food upon land conveniently enough, notwithstanding the length of their necks; for example, geese graze upon commons, and can feed themselves fat upon land. Yet is there not one landbird, which hath its neck thus disproportionate to its legs; nor one water one neither, but such as are destined by Nature in such manner as we have mentioned to search and gather their food. For Nature makes not a long neck to no purpose.

Lastly, another argument of providence and counsel relating to animals, is the various kinds of voices the same animal uses on divers occasions, and to different purposes. Hen-birds, for example, have a particular sort of voice, when they would call the male; which is so eminent in quails, that it is taken notice of by men, who by counterfeiting this voice with a quail-pipe, easily drew the cocks into their snares. The common hen all the while she is broody sits, and leads her chickens, uses a voice which we call clucking. Another she employs, when she calls her chickens to partake of any food she hath found for them; upon hearing whereof they speedily run to her. Another when upon sight of a bird of prey, or apprehension of any danger, she would scare them, bidding them, as it were, to shift for themselves; whereupon they speedily run away, and seek shelter among bushes, or in the thick grass, or elsewhere dispersing themselves far and wide. These actions do indeed necessarily infer knowledge and intention of, and direction to, the ends and uses to which they serve; not in the birds themselves, but in a superior agent, who hath put an instinct in them of using such a voice upon such an occasion; and in the young of doing that upon hearing of it, which by Providence was intended. Other voices she hath when angry, when she hath laid an egg, when in pain, or great fear, all significant, which may more easily be accounted for, as being the effects of the several passions of anger, grief, fear, joy: which yet are all argumentative of Providence, intending their several significations and uses.

I might also instance in quadrupeds; some of which have as great diversity of voices as hens themselves; and all of them significant: for example, that common domestic animal the cat, as is obvious to every one to observe, and therefore I shall not spend time to mention particulars.

Object. But against the uses of several bodies I have instanced in, that refer to man, it may be objected, that these uses were not designed by nature in the formation of the things; but that the things were by the wit of man accommodated to those uses.

To which I answer with Dr. More, in the Appendix to his 'Antidote against Atheism,' that the several useful dependencies of this kind (viz. of stones, timber, and metals, for building of houses or ships, the magnet, for navigation, &c. fire for melting of metals, and forging of instruments for the purposes, mentioned) we only find, not make them. For whether we think of of it or no, it is, for example, manifest, that fuel is good to continue fire, and fire to melt metals, and metals to make instruments to build ships and houses, and so on. Wherefore it being true, that there is such a subordinate usefulness in the things themselves that are made to our hand, it is but reason in us to impute it to such a cause, as was aware of the usefulness and serviceableness of its own works. To which I shall add, that since we find materials so fit to serve all the necessities and conveniences, and to exercise and employ the wit and industry of an intelligent and active being, and since there is such an one created that is endued with skill and ability to use them, and which by their help is enabled to rule over and subdue all inferior creatures, but without them had been left necessitous, helpless, and obnoxious to injuries above any other; and since the omniscient Creator could not but know all the uses, to which they might and would be employed by man, to them that acknowledge the being of a Deity, it is little less than a demonstration, that they were created intentionally, I do not say only, for those uses.

Methinks by all this provision for the use and service of man, the Almighty interpretatively speaks to him in this manner: I have now placed thee in a spacious and well furnished world, I have endued thee with an ability of understanding what is beautiful and proportionable, and have made that which is so, agreeable and delightful to thee; I have provided thee with materials whereon to exercise and employ thy art and strength; I have given thee an excellent instrument, the hand, accommodated to make use of them all; I have distinguished the earth into hills and valleys, and plains, and meadows, and woods; all these parts capable of culture and improvement by thy industry; I have committed to thee for thy assistance in thy labours of plowing, and carrying, and drawing, and travel; the laborious ox, the patient ass, and the strong and serviceable horse; I have created a multitude of seeds for thee to make choice out of them, of what is most pleasant to thy taste, and of most wholesome and plentiful nourishment; I have also made great variety of trees, bearing fruit both for food and physic, those too capable of being meliorated and improved by transplantation, stercoration, incision, pruning, watering, and other arts and devices. Till and manure thy fields, sow them with thy seeds, extirpate noxious and unprofitable herbs, guard them from the invasions and spoil of beasts, clear and fence in thy meadows and pastures; dress and prune thy vines, and so rank and dispose them as is most suitable to the climate; plant thee orchards, with all sorts of fruit-trees, in such order as may be most beautiful to the eye, and most comprehensive of plants; gardens for culinary herbs, and all kinds of sallading; for delectable flowers, to gratify the eye with their agreeable colours and figures, and thy scent with their fragrant odours; for odoriferous and ever-green shrubs and suffrutices; for exotic and medicinal plants of all sorts, and dispose them in that comely order, as may be most pleasant to behold, and commodious for access. I have furnished thee with all materials for building, as stone, and timber, and slate, and lime, and clay, and earth, whereof to make bricks and tiles. Deck and bespangle the country with houses and villages convenient for thy habitation, provided with out-houses and stables for the harbouring and shelter of thy cattle, with barns and granaries for the reception, and custody, and storing up thy corn and fruits. I have made thee a sociable creature, Ζῶον πολιτικὸν, for the improvement of thy understanding by conference, and communication of observations and experiments; for mutual help, assistance, and defence, build thee large towns and cities with straight and well-paved streets, and elegant rows of houses, adorned with magnificent temples for my honour and worship, with beautiful palaces for thy princes and grandees, with stately halls for public meetings of the citizens and their several companies, and the sessions of the courts of judicature, besides public porticoes and aqueducts. I have implanted in thy nature a desire of seeing strange and foreign, and finding out unknown countries, for the improvement and advance of thy knowledge in geography, by observing the bays, and creeks, and havens, and promontories, the outlets of rivers, the situation of the maritime towns and cities, the longitude and latitude, &c. of those places: in politics, by noting their government, their manners, laws, and customs, their diet and medicine, their trades and manufactures, their houses and buildings, their exercises and sports, &c. In physiology, or natural history, by searching out their natural rarities, the productions both of land and water, what species of animals, plants, and minerals, of fruits and drugs are to be found there, what commodities for bartering and permutation, whereby thou mayest be enabled to make large additions to natural history, to advance those other sciences, and to benefit and enrich thy country by increase of its trade and merchandize. I have given thee timber and iron to build the hulls of ships, tall trees for masts, flax and hemp for sails, cables and cordage for rigging. I have armed thee with courage and hardiness to attempt the seas, and traverse the spacious plains of that liquid element; I have assisted thee with a compass, to direct thy

IN THE CREATION.   143

course when thou shalt be out of all ken of land, and have nothing in view but sky and water. Go thither for the purposes before-mentioned, and bring home what may be useful and beneficial thy country in general, or thyself in particular.

I persuade myself, that the bountiful and gracious Author of man's being and faculties, and all things else, delights in the beauty of his creation, and is well pleased with the industry of man, in adorning the earth with beautiful cities and castles, with pleasant villages and country-houses, with regular gardens, and orchards, and plantations of all sorts of shrubs, and herbs and fruits, for meat, medicine, or moderate delight, with shady woods and groves, and walks set with rows of elegant trees; with pastures clothed with flocks, and valleys covered over with corn, and meadows burthened with grass, and whatever else differenceth a civil and well cultivated region, from a barren and desolate wilderness.

If a country thus planted and adorned, thus polished and civilized, thus improved to the height by all manner of culture for the support and sustenance, and convenient entertainment of innumerable multitudes of people, be not to be preferred before a barbarous and inhospitable Scythia, without houses, without plantations, without corn-fields or vineyards; where the roving hordes of the savage and truculent inhabitants, transfer themselves from place to place in waggons, as they can find pasture and forage for their cattle, and live upon milk, and flesh roasted in the sun, at the pommels of their saddles; or a rude and unpolished America, peopled with slothful and naked Indians, instead of well-built houses, living in pitiful huts and cabins, made of poles set end-ways; then surely the brute beast's condition and manner of living, to which, what we have mentioned doth nearly approach, is to be esteemed better than man's, and wit and reason was in vain bestowed on him.

Lastly, I might draw an argument of the admirable art and skill of the Creator and composer of them, from the incredible smallness of some of those natural and enlivend machines, the bodies of animals.

Any work of art of extraordinary fineness and subtlety, be it but a small engine or movement, or a curious carved or turned work of ivory or metals, such as those cups turned of ivory by Oswaldus Nerlinger of Suevia, mentioned by Joan. Faber in his expositions of Recchus's Mexican animals, which all had the perfect form of cups, and were gilt with a golden border about the brim, of that wonderful smallness, that Faber himself put a thousand of them into an excavated pepper-corn; and when he was weary of the work, and yet had not filled the vessel, his friend, John Carolus Schad, that shewed them him, put in four hundred more. Any such work, I say, is beheld with admiration, and purchased at a great rate, and treasured up as a singular rarity in the museums and cabinets of the curious, and as such, is one of the first things shewed to travellers and strangers. But what are these for their fineness and parvity (for which alone, and their figure, they are considerable), to those minute machines endued with life and motion, I mean the bodies of those animalcula, not long since discovered in pepper-water, by Mr. Leuenhoek, of Delft in Holland (whose observations were confirmed and improved by our learned and worthy countryman Dr. Robert Hook), who tells us, that some of his friends (whose testimonials he desired) did affirm, that they had seen 10,000, others 30,000, others 45,000, little living creatures, in a quantity of water no bigger than a grain of millet. And yet he made it his request to them, that they would only justify (that they might be within compass) half the number that they believed each of them saw in the water. From the greatest of these numbers he infers, that there will be 8,280,000 of these living creatures seen in one drop of water; which number (saith he) I can with truth affirm I have discerned. This (proceeds he) doth exceed belief. But I do affirm, if a larger grain of sand were broken into 8,000,000, of equal parts, one of these would not exceed the bigness of one of those creatures. Dr. Hook tells us, that after he had discovered vast multitudes of those exceeding small creatures which Mr. Leuenhoek had described, upon making use of other lights and glasses, he not only magnified those he had discovered to a very great bigness, but discovered many other sorts very much smaller than them he first saw, and some of them so exceeding small, that millions of millions might be contained in one drop of water. If Pliny, considering such insects as were known to him, and those were none but what were visible to the naked eye, was moved to cry out, that the artifice of nature was nowhere more conspicuous than in these. And again, 'In his tam parvis atque tam nullis quæ ratio, quanta vis, quam inextricabilis perfectio?' and again, 'Rerum natura nusquam magis quam in minimis tota, est,' Hist. Nat. l. ii. c. 1. What would he have said, if he had seen animals of so stupendous smallness, as I have mentioned? How would he have been rapt into an ecstasy of astonishment and admiration?

Again: if considering the body of a gnat (which by his own confession is none of the least of insects), he could make so many admiring queries, where hath nature disposed of so many senses in a gnat? ' Ubi visum in eo prætendit? ubi gustatum applicavit? ubi odoratum inseruit? ubi vero truculentam illam, et portione maximam vocem ingeneravit? qua subtilitate pennas adnexuit? prælongavit pedum crura? disposuitjejunam caveam, uti alvum? avidam sanguinis, et potissimum humani, sitim accendit? telum vero perfodiendo tergori quo spiculavit ingenio? atque ut in capaci, cum cerni non possit exilitas, ita reciproca generavit arte, ut fidiendo acuminatum pariter sorbendoque fistulosum esset.' Which words should I translate, would lose of their emphasis and elegancy. If, I say, he could make such queries about the members of a gnat, what may we make? and what would he in all likelihood have made, had he seen these incredible small living creatures? How would he have admired the immense subtility (as he phrases it) of their parts? for, to use Dr. Hook's words in his Microscopium, p. 103. if these creatures be so exceeding small, what must we think of their muscles and other parts? certain it is, that the mechanism by which nature performs the muscular motion, is exceeding small and curious; and to the performance of every muscular motion, in greater animals at least, there are not fewer distinct parts concerned than many millions of millions, and these visible through a microscope.

Use.—Let us then consider the works of God, and observe the operations of his hands: let us take notice of, and admire his infinite wisdom and goodness in the formation of them: no creature in this sublunary world is capable of so doing, beside man; and yet we are deficient herein: we content ourselves with the knowledge of the tongues, and a little skill in philology, or history perhaps, and antiquity, and neglect that which to me seems more material, I mean natural history, and the works of the creation. I do not discommend, or derogate from those other studies: I should betray mine own ignorance and weakness should I do so; I only wish they might not altogether justle out, and exclude this. I wish that this might be brought in fashion among us; I wish men would be so equal and civil, as not to disparage, deride, and vilify those studies which themselves skill not of, or are not conversant in; no knowledge can be more pleasant than this, none that doth so satisfy and feed the soul; in comparison whereto that of words and phrases seems to me insipid and jejune. That learning (saith a wise and observant prelate) which consists only in the form and pedagogy of arts, or the critical notion upon words and phrases, hath in it this intrinsical imperfection, that it is only so far to be esteemed, as it conduceth to the knowledge of things, being in itself but a kind of pedantry, apt to infect a man with such odd humours of pride, and and affectation, and curiosity, as will render him unfit for any great employment. Words being but the images of matter, to be wholly given up to the study of these, what is it but Pygmalion's frenzy, to fall in love with a picture or image. As for oratory, which is the best skill about words, that hath by some wise men been esteemed but a voluntary art, like to cookery, which spoils wholsome meats, and helps unwholsome, by the variety of sauces, serving more to the pleasure of taste, than the health of the body.

It may be (for aught I know, and as some divines have thought) part of our business and employment in eternity, to contemplate the works of God, and give him the glory of his wisdom, power, and goodness manifested in the creation of them. I am sure it is part of the business of a sabbath-day, and the sabbath is a type of that eternal rest; for the sabbath seems to have been first instituted for a commemoration of the works of the creation, from which God is said to have rested upon the seventh day.

It is not likely that eternal life shall be a torpid and unactive state, or that it shall consist only in an uninterrrupted and endless act of love; the other faculties shall be employed as well as the will, in actions suitable to, and perfective of their natures; especially the understanding, the supreme faculty of the soul, which chiefly differenceth us from brute beasts, and makes us capable of virtue and vice, of rewards and punishments, shall be busied and employed in contemplating the works of God, and observing the divine art and wisdom, manifested in the structure and composition of them; and reflecting upon their great Architect the praise and glory due to him. Then shall we clearly see to our great satisfaction and admiration, the ends and uses of these things, which here were either too subtile for us to penetrate and discover, or too remote and unaccessible for us to come to any distinct view of, viz. the planets and fixed stars; those illustrious bodies, whose contents and inhabitants, whose stores and furniture we have here so longing a desire to know, as also their mutual subserviency to each other. Now the mind of man being not capable at once to advert to more than one thing, a particular view and examination of such an innumerable number of vast bodies, and the great multitude of species, both of animate and inanimate beings, which each of them contains, will afford matter enough to exercise and employ our minds, I do not say, to all eternity, but to many ages, should we do nothing else.

Let it not suffice us to be book-learned, to read what others have written, and to take upon trust more falsehood than truth: but let us ourselves examine things as we have opportunity, and converse with nature as well as books. Let us endeavour to promote and increase this knowledge, and make new discoveries, not so much distrusting our own parts, or despairing of our own abilities, as to think that our industry can add nothing to the inventions of our ancestors, or correct any of their mistakes. Let us not think that the bounds of science are fixed like Hercules's pillars, and inscribed with a *Ne plus ultra*. Let us not think we have done, when we have learnt what they have delivered to us. The treasures of nature are inexhaustible. Here is employment enough for the vastest parts, the most indefatigable industries, the happiest opportunities, the most prolix and undisturbed vacancies. 'Multa venientis ævi populus ignota nobis sciet: multa sæculis tunc futuris, cum memoria nostri exoleverit, reservantur. Pusilla res mundus est, nisi in eo quod quærat omnis mundus habeat,' Seneca Nat. Quæst. lib. 7. cap. 31. 'The people of the next age shall know many things unknown to us: many are reserved for ages then to come, when we shall be quite forgotten, no memory of us remaining. The world would be a pitiful small thing indeed, if it did not contain enough for the inquiries of the whole world.' Yet, and again, Epist. 64. 'Multum adhuc restat operis, multumque restabit, nec ulli nato post mille sæcula præcludetur occasio aliquid adhuc adjiciendi.' ' Much work still remains, and much will remain, neither to him that shall be born after a thousand ages, will matter be wanting for new additions to what hath already been invented.' Much might be done, would we but endeavour, and nothing is insuperable to pains and patience. I know that a new study, at first, seems very vast, intricate, and difficult; but after a little resolution and progress, after a man becomes a little acquainted, as I may so say, with it, his understanding is wonderfully cleared up and enlarged, the difficulties vanish, and the thing grows easy and familiar. And for our encouragement in this study, observe what the Psalmist saith, Psalm. iii. 1. ' The works of the Lord are great, sought out of all them that have pleasure therein.' Which though it be principally spoken of the works of Providence, yet may as well be verified of the works of creation. I am sorry to see so little account made of real experimental philosophy in this university;[*] and that those ingenious sciences of the mathematics are so much neglected by us: and therefore do earnestly exhort those that are young, especially gentlemen, to set upon these studies, and take some pains in them. They may possibly invent something of eminent use and advantage to the world; and one such discovery would abundantly compensate the expense and travail of one man's whole life. However, it is enough to maintain and continue what is already invented: neither do I see what more ingenious and manly employment they can pursue, tending more to the satisfaction of their own minds, and the illustra-

[*] Cambridge, where the author lived, at the first writing of this.

tion of the glory of God. For he is wonderful in all his works.

But I would not have any man cross his natural genius or inclinations, or undertake such methods of study, as his parts are not fitted to, or not serve those ends to which his friends upon mature deliberation have designed him; but those who d abound with leisure, or who have a natural propension and genius inclining them thereto, or those who by reason of the strength and greatness of their parts, are able to compass and comprehend the whole latitude of learning.

Neither yet need those who are designed to divinity itself, fear to look into these studies, or think they will engross their whole time, and that no considerable progress can be made therein, unless men lay aside and neglect their ordinary callings, and necessary employments. No such matter; our life is long enough, and we might find time enough, did we husband it well: 'Vitam non accepimus brevem sed fecimus, nec inopes ejus sed prodigi sumus,' as Seneca saith; 'we have not received a short life, but have made it so; neither do we want time, but are prodigal of it.' And did the young men fill up that time with these studies, which lie upon their hands, which they are incumbered with, and troubled how to pass away, much might be done even so. I do not see, but the study of true physiology may be justly accounted a proper Προπαιδεία, or preparative to divinity. But to leave that, it it is a generally received opinion, that all this visible world was created for man; that man is the end of the creation, as if there were no other end of any creature, but some way or other to be serviceable to man. This opinion is as old as Tully; for, saith he, in his second book, De Nat. Deorum, ' Principio ipse mundus deorum hominumque causa factus est; quæque in eo sunt omnia, ea parata ad fructum hominum, et inventa sunt.' But though this be vulgarly received, yet wise men now a-days think otherwise. Dr. More* affirms, 'That creatures are made to enjoy themselves, as well as to serve us: and that it is a gross piece of ignorance and rusticity to think otherwise.' And in another place: 'This comes only out of pride and ignorance, or a haughty presumption, because we are encouraged to believe, that, in some sense, all things are made for man, therefore to think that they are not at all made for themselves. But he that pronounceth this, is ignorant of the nature of man, and the knowledge of things: for if a good man be merciful to his beast, then surely, a good God is bounand benign, and takes pleasure that all his creatures enjoy themselves, that have life and sense, and are capable of enjoyment.'

Those philosophers, indeed, who hold man to be the only creature in this sublunary world, endued with sense and perception, and that all other animals are mere machines or puppets, have some reason to think that all things here below were made for man. But this opinion seems to me too mean, and unworthy the majesty, wisdom, and power of God; nor can it well consist with his veracity, instead of a multitude of noble creatures, endued with life and sense, and spontaneous motion, as all mankind until of late years believed, and none ever doubted of (so that it seems we are naturally made to think so), to have stocked the earth with divers sets of automata, without all sense and perception, being wholly acted from without, by the impulse of external objects.

But be this so, there are infinite other crea-

* Antid. Atheism. l. ii. c. 11.

tures without this earth, which no considerate man can think were made only for man, and have no other use. For my part, I cannot believe that all the things in the world were so made for man, that they have no other use.

For it seems to me highly absurd and unreasonable, to think that bodies of such vast magnitude as the fixed stars, were only made to twinkle to us; nay, a multitude of them there are, that do not so much as twinkle, being either, by reason of their distance or of their smallness, altogether invisible to the naked eye, and only discoverable by a telescope; and it is likely perfecter telescopes than we yet have, may bring to light many more; and who knows how many lie out of the ken of the best telescope that can possibly be made? And I believe there are many species in nature, even in this sublunary world, which were never yet taken notice of by man, and consequently of no use to him, which yet we are not to think were created in vain; but may be found out by, and of use to those who shall live after us in future ages. But though in this sense it be not true that all things were made for man, yet thus far it is, that all the creatures in the world may be some way or other useful to us, at least to exercise our wits and understandings, in considering and contemplating of them, and so afford us subject of admiring and glorifying their and our Maker. Seeing then, we do believe, and assert, that all things were in some sense made for us, we are thereby obliged to make use of them for those purposes for which they serve us, else we frustrate this end of their creation. Now some of them serve only to exercise our minds: many others there be, which might probably serve us to good purpose, whose uses are not discovered, nor are they ever like to be, without pains and industry. True it is, many of the greatest inventions have been accidentally stumbled upon, but not by men supine and careless, but busy and inquisitive. Some reproach methinks it is to learned men, that there should be so many animals still in the world, whose outward shape is not yet taken notice of or described, much less their way of generation, food, manners, uses, observed.

The Scripture, Psalm cxlviii. calls upon the sun, moon, and stars; fire and hail, snow and vapour; stormy winds and tempests, mountains and all hills; fruitful trees and all cedars; beasts and all cattle; creeping things and flying fowl, &c. 'to praise the Lord.' How can that be? Can senseless and inanimate things praise God? Such as are the sun, and moon, and stars. And although beasts be advanced higher to some degree of sense and perception; yet being void of reason and understanding, they know nothing of things, or of the Author and Maker of themselves, and other creatures. All that they are capable of doing, in reference to the praising of God, is (as I said before) by affording matter or subject of praising him, to rational and intelligent beings. So the Psalmist, Psal. xix. 1. 'The heavens declare the glory of God, and the firmament sheweth his handy-work.' And therefore, the Psalmist, when he calls upon sun, and moon, and stars, to praise God, doth in effect call upon men and angels, and other rational beings, to consider those great effects of the Divine power and wisdom, their vast dimensions, their regular motions and periods, their admirable disposition and order, their eminent ends and uses in illuminating and enlivening the planets, and other bodies about them, and their inhabitants, by their comfortable and cherishing light, heat, and influences, and to give God the glory of his power, in making such great and illustrious bodies, and of his wisdom and goodness in so placing and disposing of them, so moving them regularly and constantly, without clashing or interfering one with another, and enduing them with such excellent virtues and properties as to render them so serviceable and beneficial to man, and all other creatures about them.

The like may be said of fire, hail, snow, and other elements and meteors; of trees, and other vegetables; of beasts, birds, insects, and all animals, when they are commanded to praise God, which they cannot do by themselves. Man is commanded to consider them particularly, to observe and take notice of their curious structure, ends, and uses, and give God the praise of his wisdom, and other attributes therein manifested.

And therefore those who have leisure, opportunity, and abilities, to contemplate and consider any of these creatures, if they do it not, do as it were rob God of some part of his glory, in neglecting or slighting so eminent a subject of it, and wherein they might have discovered so much art, wisdom, and contrivance.

And it is particularly remarkable, that the divine author of this psalm, amongst other creatures, calls upon insects also to praise God; which is as much as to say, Ye sons of men, neglect none of his works, those which seem most vile and contemptible; there is praise belongs to him for them. Think not that any thing he hath vouchsafed to create, is unworthy thy cognisance, to be slighted by thee. It is pride and arrogance, or ignorance and folly in thee so to think. There is a greater depth of art and skill in the structure of the meanest insect, than thou art able for to fathom or comprehend.

The wisdom, art, and power of Almighty God,

shines forth as visibly in the structure of the body of the minutest insect, as in that of a horse or elephant: therefore God is said to be, *Maximus in minimis.* We men, esteem it a more difficult matter, and of greater art and curiosity to frame a small watch, than a large clock: and no man blames him who spent his whole time in the consideration of the nature and works of a bee, or thinks his subject was too narrow. Let us not then esteem any thing contemptible or inconsiderable, or below our notice taking; for this is to derogate from the wisdom and art of the Creator, and to confess ourselves unworthy of those endowments of knowledge and understanding which he hath bestowed on us. Do we praise Dædalus, and Architas, and Hero, and Callicrates, and Albertus Magnus, and many others which I might mention, for their cunning in inventing, and dexterity in framing and composing a few dead engines or movements, and shall we not admire and magnify the great Δημιουργὸς Κόσμου, Former of the world? Who hath made so many; yea, I may say, innumerable, rare pieces, and those too not dead ones, such as cease presently to move so soon as the spring is down, but all living, and themselves performing their own motions, and those so intricate and various, and requiring such a multitude of parts and subordinate machines, that it is incomprehensible, what art, and skill, and industry, must be employed in the framing of one of them.

I have already noted out of Dr. Hook, that to the performance of every muscular motion, at least in greater animals, there are not fewer distinct parts concerned, than many millions of millions.

Further, from the consideration of our own smallness and inconsiderableness, in respect of the greatness and splendour of those glorious heavenly bodies, the sun, moon, and stars, to which our bodies bear no proportion at all, either in magnitude or lustre; let us with the holy psalmist raise up our hearts, to magnify the goodness of God towards us, in taking such notice of us, and making such provision for us, and advancing us so highly above his works, Psal. viii. 3. 'When I consider the heavens the work of thy fingers, the moon and the stars which thou hast ordained. What is man that thou art mindful of him? and the Son of Man that thou visitest him? For thou hast made him a little lower than the angels, and hast crowned him with glory and honour,' &c.

But it may be objected, that God Almighty was not so selfish and desirous of glory, as to make the world and all the creatures therein, only for his own honour, and to be praised by man. To assert this, were, in Des Cartes's opinion, an absurd and childish thing, and a resembling of God to a proud man. It is more worthy the Deity to attribute the creation of the world to the exundation and overflowing of his transcendent and infinite goodness, which is of its own nature, and in the very notion of it most free, diffusive, and communicative.

To this I shall answer in two words. First, the testimony of Scripture makes God in all his actions to intend and design his own glory mainly, Prov. xvi. 4. 'God made all things for himself.' How! for himself? he had no need of them; he had no use of them. No, he made them for the manifestation of his power, wisdom, and goodness, and that he might receive from the creatures that were able to take notice thereof, his tribute of praise. Psal. l. 14. 'Offer unto God thanksgiving.' And in the next verse, 'I will deliver thee, and thou shalt glorify me.' And again in the last verse 'Whoso offereth praise, glorifieth me.' So praise is called a sacrifice, and 'the calves of the lips,' Hosea xiv. 2. Isa. xlii. 8. 'I am the Lord, that is my name, and my glory will I not give to another.' Isa. xlviii. 11. 'And I will not give my glory to another.' The Scripture calls upon the heavens and earth, and sun, and moon, and stars, and all other creatures, to praise the Lord; that is, by the mouth of man (as I shewed before), who is hereby required to take notice of all these creatures, and to admire and praise the power, wisdom, and goodness of God manifested in the creation and designations of them.

Secondly, It is most reasonable that God Almighty should intend his own glory: for he being infinite in all excellences and perfections, and independent upon any other being; nothing can be said or thought of him too great, and which he may not justly challenge as his due; nay, he cannot think too highly of himself, his other attributes being adequate to his understanding; so that, though his understanding be infinite, yet he understands no more than his power can effect, because that is infinite also. And therefore it is fit and reasonable, that he should own and accept the creature's acknowledgments and celebrations of those virtues and perfections, which he hath not received of any other, but possesseth eternally and originally of himself. And indeed (with reverence be it spoken) what else can we imagine the ever-blessed Deity to delight and take complacency in for ever, but his own infinite excellencies and perfections, and the manifestations and effects of them, the works of the creation, and the sacrifices of praise and thanks offered up by such of his creatures as are capable of considering those works, and discerning the traces and footsteps of his power and wisdom appearing in the formation of them; and moreover, whose bounden duty it is so to do. The reason why man ought not to admire himself, or seek his own glory, is, because he is a dependent creature, and hath nothing but what he hath received, and not only dependent, but imperfect; yea, weak and impotent. And yet I do not take humility in man to consist in disowning or denying any gift or ability that is in him, but in a just valuation of such gifts and endowments, yet rather thinking too meanly than too highly of them; because human nature is so apt to err in running into the other extreme, to flatter itself, and to accept those praises that are not due to it; pride being an elation of spirit upon false grounds, or a desire and acceptance of undue honour. Otherwise, I do not see why a man may not admit, and accept the testimonies of others, concerning any perfection, accomplishment, or skill, that he is really possessed of; yet can he not think himself to deserve any praise or honour for it, because both the power and the habit are the gift of God: and considering that one virtue is counterbalanced by many vices; and one skill or perfection, with much ignorance and infirmity.

PART II.

I proceed now to select some particular pieces of the creation, and to consider them more distinctly. They shall be only two:

I. The whole body of the earth.
II. The body of man.

First, The body of the earth: and therein I shall take notice of, 1. Its figure. 2. Its motion. 3. The constitution of its parts.

By earth I here understand not the dry land, or the earth contradistinguished to water, or the earth considered as an element: but the whole terraqueous globe, composed of earth and water.

1. For the figure; I could easily demonstrate it to be spherical. That the water, which by reason of its fluidity should, one would think, compose itself to a level, yet doth not so, but hath a gibbose superficies, may to the eye be demonstrated upon the sea. For when two ships sailing contrary ways lose the sight one of another: first the keel and hull disappear, afterward the sails, and if when upon deck you have lost sight of all, you get up to the top of the mainmast you may descry it again. Now what should take away the sight of these ships from each other, but the gibbosity of the interjacent water? The roundness of the earth from north to south is demonstrated from the appearance of northern stars above the horizon, and loss of the southern to them that travel northward; and on the contrary the loss of the northern, and appearance of the southern to them that travel southward. For were the earth a plain, we should see exactly the same stars wherever we were placed on that plain. The roundness from east to west

is demonstrated from eclipses of either of the great luminaries. For why the same eclipse, suppose of the sun, which is seen to them that live more easterly, when the sun is elevated six degrees above the horizon, should be seen to them that live one degree more westerly, when the sun is but five degrees above the horizon, and so lower and lower proportionably to them that live more and more westerly, till at last it appear not at all, no account can be given, but the globosity of the earth. For were the earth a perfect plain, the sun would appear eclipsed to all that live upon the plain, if not exactly in the same elevation, yet pretty near it; but to be sure it would never appear to some, the sun being elevated high above the horizon; and not at all to others. It being clear then, that the figure of the earth is spherical, let us consider the conveniences of this figure.

1. No figure is so capacious as this, and consequently, whose parts are so well compacted and united, and lie so near one to another for mutual strength. Now the earth, which is the basis of all animals, and as some think of the whole creation, ought to be firm, and stable, and solid, and as much as is possible secured from all ruins and concussions.

2. This figure is most consonant and agreeable to the natural nutus, or tendency of all heavy bodies. Now the earth being such a one, and all its parts having an equal propension, or connivency to the centre, they must needs be in greatest rest, and most immoveable when they are all equidistant from it. Whereas, were it an angular body, all the angles would be vast and steep mountains, bearing a considerable proportion to the whole bulk, and therefore those parts being extremely more remote from the centre,

than those about the middle of the plains, would consequently press very strongly thitherward; and unless the earth were made of adamant and marble, in time the other parts would give way, till all were levelled.

3. Were the earth an angular body, and not round, all the whole earth would be nothing else but vast mountains, and so incommodious for animals to live upon. For the middle point of every side would be nearer the centre than any other, and consequently from that point, which way soever one travelled would be up hill, the tendency of all heavy bodies being perpendicularly to the centre. Besides, how much this would obstruct commerce is easily seen. For not only the declivity of all places would render them very difficult to be travelled over, but likewise the midst of every side being lowest and nearest the centre, if there were any rain, or any rivers, must needs be filled with a lake of water, there being no way to discharge it, and possibly the water would rise so high, as to overflow the whole latus. But surely, there would be much more danger of the inundation of whole countries than now there is: all the waters falling upon the earth, by reason of its declivity every way, easily descending down to the common receptacle the sea. And these lakes of water being far distant one from another, there could be no commerce between far remote countries, but by land.

4. A spherical figure is most commodious for dinetical motion, or revolution upon its own axis. For in that neither can the medium at all resist the motion of the body, because it stands not in its way, no part coming into any space but what the precedent left, neither doth one part of the superficies move faster than another; whereas were it angular, the parts about the angles would

find strong resistance from the air, and those parts also about the angles would move much faster than those about the middle of the plains, being remoter from the centre than they. It remains, therefore, that this figure is the most commodious for motion.

Here I cannot but take notice of the folly and stupidity of the Epicureans, who fancied the earth to be flat and contiguous to the heavens on all sides, and that it descended a great way with long roots; and that the sun was new made every morning, and not much bigger than it seems to the eye, and of a flat figure, and many other such gross absurdities as children among us would be ashamed of.

Secondly, I come now to speak of the motion of the earth. That the earth (speaking according to philosophical accurateness) doth move both upon its own poles, and in the ecliptic, is now the received opinion of the most learned and skilful mathematicians. To prove the diurnal motion of it upon its poles, I need produce no other argument than, first, the vast disproportion in respect of magnitude that is between the earth and the heavens, and the great unlikelihood that such an infinite number of vast bodies should move about so inconsiderable a spot as the earth, which in comparison with them, by the concurrent suffrages of mathematicians of both persuasions, is a mere point, that is, next to nothing. Secondly, the immense and incredible celerity of the motion of the heavenly bodies in the ancient hypothesis. Thirdly, of its annual motion in the ecliptic, the stations and retrogradations of the superior planets are a convincing argument, there being a clear and facile account thereof to be given from the mere motion of the earth in the ecliptic; whereas in the old hypothesis no account can be given, but by the unreasonable fiction of epicycles and contrary motions; add hereto the great unlikelihood of such an enormous epicycle as Venus must describe about the sun, not under the sun, as the old astronomers fancied. About the sun I say, as appears by its being hid or eclipsed by it, and by its several phases, like the moon. So that whosoever doth clearly understand both hypotheses, cannot, I persuade myself, adhere to the old and reject the new, without doing some violence to his faculties.

Against this opinion lie two objections, First, That it is contrary to sense, and the common opinion and belief of mankind. Secondly, That it seemeth contrary to some expressions in Scripture. To the first, I answer, that our senses are sometimes mistaken, and what appears to them is not always in reality so as it appears. For example, the sun or moon appear no bigger, at most, than a cartwheel, and of a flat figure. The earth seems to be plain; the heavens to cover it like a canopy, and to be contiguous to it round about. A firebrand nimbly moved round, appears like a circle of fire; and to give a parallel instance, a boat lying still at anchor in a river, to him that sails or rows by it, seems to move apace: and when the clouds pass nimbly under the moon, the moon itself seems to move the contrary way. And there have been whole books written in confutation of vulgar errors. Secondly, As to the Scripture, when speaking of these things, it accommodates itself to the common and received opinions, and employs the usual phrases and forms of speech (as all wise men also do, though in strictness, they be of a different or contrary opinion), without intention of delivering any thing doctrinally concerning these points, or confuting the contrary: and yet by those that maintain the opinion of the earth's motion, there might a convenient interpretation be given of such places as seem to contradict it. Howbeit, because some pious persons may be offended at such an opinion, as savouring of novelty, thinking it inconsistent with divine revelation, I shall not positively assert it, only propose it as an hypothesis not altogether improbable. Supposing, then, that the earth doth move, both upon its own poles, and in the ecliptic about the sun, I shall shew how admirably its situation and motion are contrived for the conveniency of man and other animals: which I cannot do more fully and clearly than Dr. More hath already done in his 'Antidote against Atheism,' whose words therefore I shall borrow.

First, Speaking of the parallelism of the axis of the earth, he saith, I demand whether it be better to have the axis of the earth steady and perpetually parallel to itself, or to have it carelessly tumble this way and that way as it happens, or at least very variously and intricately? And you cannot but answer me, it is better to have it steady and parallel. For in this lies the necessary foundation of the art of navigation and dialling. For that steady stream of particles, which is supposed to keep the axis of the earth parallel to itself, affords the mariner both his cynosura, and his compass. The loadstone and the loadstar depend both upon this. The loadstone, as I could demonstrate, were it not too great a digression; and the loadstar, because that which keeps the axis parallel to itself, makes each of the poles constantly respect such a point in the heavens; as for example, the north pole to point almost directly to that which

H

we call the pole-star. And besides, dialling could not be at all without this steadiness of the axis. But both these arts are pleasant, and one especially of mighty importance to mankind: for thus there is an orderly measuring of our time for affairs at home, and an opportunity of traffic abroad with the most remote nations of the world, and so there is a mutual supply of the several commodities of all countries, besides the enlarging our understandings by so ample experience we get ooth of men and things. Wherefore if we were rationally to consult, whether the axis of the earth were better be held steady and parallel to itself, or left at random, we would conclude it ought to be steady, and so we find it 'de faito,' though the earth move floating in the liquid heavens. So that, appealing to our own faculties, we are to affirm, that the constant direction of the axis of the earth was established by a principle of wisdom and counsel.

Again, There being several postures of this steady direction of the axis of the earth, viz. Either perpendicular to a plane, going through the centre of the sun, or coincident, or inclining, I demand which of all these reason and knowledge would make choice of? Not of a perpendicular posture; for so both the pleasant variety, and great convenience of summer and winter, spring and autumn, would be lost, and for want of accession of the sun, these parts of the earth which now bring forth fruits, and are habitable, would be in an incapacity of ever bringing forth any; sith then, the heat could never be greater than now it is at our 10th of March, or the 11th of September, and therefore not sufficient to bring their fruits and grain to maturity, and consequently could entertain no inhabitants; and those parts that the full heat of the sun could reach, he plying them always alike without any annual recession or intermission, would at last grow tired or exhausted, to be wholly dried up, and want moisture, the sun dissipating and casting off the clouds northwards and southwards. Besides, we observe than an orderly vicissitude of things doth much more gratify the contemplative property in man.

And now, in the second place, neither would reason make choice of a coincident position. For if the axis thus lay in a plane that goeth through the centre of the sun, the ecliptic would, like a colure, or one of the meridians, pass through the poles of the earth, which would put the inhabitants of the world into a pitiful condition. For they that escape best in the temperate zone, would be accloyed with long nights very tedious, no less than forty days; and those that now never have their night above twenty-four hours, as Friesland, Iceland, the farthest parts of Russia and Norway, would be deprived of the sun above a hundred and thirty days together. Ourselves in England, and the rest of the same clime, would be closed up in darkness no less than a hundred or eighty days; and so proportionably of the rest, both in and out of the temperate zones. And as for summer and winter, though those vicissitudes would be, yet it could not but cause raging diseases, to have the sun stay so long, describing his little circles so near the poles, and lying so hot on the inhabitants, that had been in so long extremity of darkness and cold before.

It remains, therefore, that the posture of the axis of the earth be inclining, not perpendicular, not coincident to the fore-mentioned plain. And verily, it is not only inclining, but in so fit a proportion, that there can be no fitter imagined to make it, to the utmost capacity, as well pleasant as habitable. For though the course of the sun be curbed between the tropics, yet are not those parts directly subject to his perpendicular beams, either unhabitable, or extremely hot, as the ancients fancied: by the testimony of travellers, and particularly Sir Walter Raleigh, the parts under and near the line being as fruitful and pleasant, and fit to make a paradise of, as any in the world. And that they are as suitable to the nature of man, and as convenient to live in, appears from the longevity of the natives; as, for instance, the Ethiops, called by the ancients, Μακρόβιοι; but especially in the Brasilians in America, the ordinary term of whose life is a hundred years, as is set down by Piso, a learned physician of Holland, who travelled thither on purpose to augment natural knowledge, but especially what related to physic. And reasonable it is, that this should be so, for neither doth the sun lie long upon them, their day being but twelve hours, and their night as long, to cool and refresh them, and besides they have frequent showers, and constant breezes, or fresh gales of wind from the east. It was the opinion of Asclepiades, as Plutarch reports, that generally the inhabitants of cold countries are longer lived than those of hot, because the cold keeps in the natural heat, as it were locking up the pores to prevent its evaporation; whereas in hot regions the heat is easily dissipated, the pores being large and open to give it way. Which opinion because I find some learned men still to adhere to, I shall produce some farther instances out of Monsieur Rochefort's history of the Antilles Islands, to confirm the contrary, and to shew how often and easily we may be deceived, if we trust to our own ratiocinations, how plausible soever, and consult not experience.

The ordinary life, saith he, of our Caribbeans is a hundred and fifty years long, and sometimes more. There were some among them, not long since living, who remembered to have seen the first Spaniards that aborded America, who, we may thence conclude, lived to be at least 160 years old.

The Hollanders who traffic in the Molucca Islands, assure us, that the ordinary term of life of the natives there is one hundred and thirty years.

Vincent le Blanc tells us, that in Sumatra, Java, and the neighbouring islands, the life of the inhabitants is extended to 140 years, and that in the realm of Cassuby it reaches 150. Francis Pirard promotes the life of the Brasilians beyond the term we have set it, v. g. to 160 years or more, and says that in Florida and Jucatan there are men found who pass that age. And it is said, that the French in Laudonier's voyage into Florida, anno 1564, saw a certain old man, who affirmed himself to be three hundred years old, and the father of five generations; and well he might be of double that number.

Lastly, Mapheus reports, that a certain Bengalese vaunted himself to be 335 years old. So far Monsieur Rochefort. Indeed, these two last instances, being perchance singular and extraordinary, do not prove the point; for even among us, where the ordinary term of life is about threescore and ten, or fourscore, there occur some rare instances of persons who have lived 130, 140, 150 years, and more. But the other testimonies being general, prove it beyond contradiction. Neither yet is the thing in itself improbable: for there being not so great inequality of weather in those hot countries, as there is in cold, the body is kept in a more equal temper, and, not having such frequent shocks, as are occasioned by such air, and often changes, and that from one extreme to another, holds out much longer. So we see infirm and crazy persons, when they come to be so weak as to be fixed to their beds, hold out many years, some I have heard of, that have lain bed-rid twenty years : because in their bed they are always kept in an almost equal temper of heat, who, had they been exposed to the excesses of heat and cold, would not probably have survived one.

Seeing, then, this best posture which our reason could make choice of, we see really established in nature, we cannot but acknowledge it to be the issue of wisdom, counsel, and providence. Moreover, a farther argument to evince this is, that though it cannot but be acknowledged, that if the axis of the earth were perpendicular to the plane of the ecliptic, her motion would be more easy and natural, yet notwithstanding, for the conveniences fore-mentioned, we see it is made in an inclining posture.

Another very considerable, and heretofore unobserved convenience of this inclination of the earth's axis, Mr. Kiel affords us in his 'Examination of Dr. Burnet's Theory of the Earth,' p. 69.

There is, saith he, one more (besides what he has mentioned before) considerable advantage, which we reap by the present position of the earth, which I will here insert, because I do not know that it is taken notice of by any. And it is, that by the present inclination of the earth's axis to the plane of the ecliptic, we who live beyond 45 degrees of latitude, and stand most in need of it, have more of the sun's heat throughout the year, than if he had shined always in the equator; that is, if we take the sum of the sun's actions upon us both in summer and winter, they are greater than its heat would be if he moved always in the equator; or, which is the same thing, the aggregate of the sun's heat upon us while he describes any two opposite parallels, is greater than it would be if in those two days he described the equator. Whereas in the torrid zone, and even in the temperate almost as far as 45 degrees of latitude, the sun of the sun's heat in summer and winter is less than it would be, were the axis of the earth perpendicular to the plane of the ecliptic. For the demonstration of which, I refer the reader to the book itself.

I think, proceeds he, this consideration cannot but lead us into a transcendent admiration of the divine wisdom, which hath placed the earth in such a posture as brings with it several conveniences beyond what we can easily discover without study and application: and I make no question, but if the rest of the works of nature were well observed, we should find several advantages which accrue to us by their present constitution, which are far beyond the uses of them that are yet discovered; by which it will plainly appear, that God hath chosen better for us than we could have done for ourselves.

If any man should object and say, it would be more convenient for the inhabitants of the earth, if the tropics stood at a greater distance, and the sun moved farther northward and southward, for so the north and south parts would be relieved, and not exposed to so extreme cold, and thereby rendered unhabitable as now they are.

To this I answer, that this would be more inconvenient to the inhabitants of the earth in general, and yet would afford the north and south parts but little more comfort. For then as much as the distances between the tropics were enlarged, so much would also the arctic and antarctic circles be enlarged too; and so we in England, and so on northerly, should not have that grateful and useful succession of day and night; but proportionably to the sun's coming towards us, so would our days be of more than twenty-four hours' length, and according to his recess in winter our nights proportionably, which how great an inconvenience it would be, is easily seen. Whereas now the whole latitude of earth, which hath at any time above twenty-four hours' day, and twenty-four hours' night, is little and inconsiderable in comparison of the whole bulk, as lying near the poles. And yet neither is that part altogether unuseful, for in the waters there live fishes, which otherwhere are not obvious; so we know the chief whale-fishing is in Greenland; yea, not only fish, but great variety of water-fowl, both whole and cloven-footed, frequent the waters and feed there, breeding also on the cliffs by the sea-side, as they do with us. The figures and descriptions of a great many whereof are given us by Martin in his voyage to Spitzberg, or Greenland: and on the land, bears, and foxes, and deer, in the most northerly country that was ever yet touched, and doubtless, if we shall discover farther to the very north pole, we shall find all that tract not to be vain, useless, or unoccupied.

Thirdly, The third and last thing I proposed, was the constitution and consistency of the parts of the earth. And first, admirable it is, that the waters should be gathered together into such great conceptacula, and the dry land appear; and though we had not been assured thereof by divine revelation, we could not in reason but have thought such a division and separation to have been the work of omnipotency and infinite wisdom and goodness. For in this condition the water nourishes and maintains innumerable multitudes of various kinds of fishes; and the dry land supports and feeds as great variety of plants and animals, which have there firm footing and habitation. Whereas had all been earth, all the species of fishes had been lost, and all those commodities which the water affords us; or all water, there had been no living for plants, or terrestrial animals, or man himself, and all the beauty, glory, and variety of this inferior world had been gone, nothing being to be seen, but one uniform dark body of water: or had all been mixt and made up of water and earth into one body of mud or mire, as one would think should be most natural: for why such a separation, as at present we find, should be made, no account can be given, but Providence. I say, had all this globe been mire or mud, then could there have been no possibility for any animals at all to have lived, excepting some few, and those very dull and inferior ones too. That therefore the earth should be made thus, and not only so, but with so great variety of parts, as mountains, plains, vallies, sand, gravel, lime, stone, clay, marble, argilla, &c. which are so delectable and pleasant, and likewise so useful and convenient for the breeding and living of various plants and animals; some affecting mountains, some plains, some valleys, some watery places, some shade, some sun, some clay, some sand, some gravel, &c. That the earth should be so figured as to have mountains in the mid-land parts; abounding with springs of water pouring down streams and rivers for the necessities and conveniences of the inhabitants of the lower countries; and that the levels and plains should be formed with so easy a declivity as to cast off the water, and yet not render travelling or tillage very difficult or laborious. These things, I say, must needs be the result of counsel, wisdom, and design. Especially when (as I said before) not that way which seems more facile and obvious to chance is chosen, but that which is more difficult and hard to be traced, when it is most convenient and proper for those nobler ends and designs, which were intended by its wise Creator and Governor. Add to all this, that the whole dry land is, for the most part, covered over with a lovely carpet of green grass and other herbs, of a colour, not only most grateful and agreeable, but most useful and salutary to the eye; and this also decked and adorned with great variety of flowers of beautiful colours and figures, and of most pleasant and fragrant odours for the refreshment of our spirits, and our innocent delight: as also with beautiful shrubs, and stately trees, affording us not only pleasant and nourishing fruits, many liquors, drugs, and good medicines, but timber, and utensils for all sorts of trades, and the conveniences of man. Out of many thousands of which, we will only just name a few, lest we should be tedious, and too bulky.

First, the cocoa, or coker-nut tree, that supplies the Indians with almost whatever they stand in need of, as bread, water, wine, vinegar, brandy, milk, oil, honey, sugar, needles, thread, linen, clothes, cups, spoons, beesoms, baskets, paper, masts for ships, sails, cordage, nails, coverings for their houses, &c. Which may be seen at large in the many printed relations of voyages and travels to the East Indies, but most faithfully in the 'Hortus Malabaricus,' published by that immortal patron of natural learning, Henry

Van Rheede van Draakenstein. who has had great commands and employ in the Dutch colonies.

Secondly, The *Aloe Muricata vel Aculeata*, which yields the Americans every thing their necessities require, as fences and houses, darts, weapons, and other arms, shoes, linen and clothes, needles and thread, wine and honey, besides many utensils, for all which Hernandes, Garcilasso de la Vega, and Margrave may be consulted.

Thirdly, The Bandura Cingalensium, called by some the Priapus Vegetabilis, at the end of whose leaves hang long sacks or bags, containing a pure limpid water, of great use to the natives when they want rain for eight or ten months together.

A parallel instance to this of the Bandura, my learned and worthy friend Dr. Sloane affords us in a plant by him observed in the island of Jamaica, and described by the title of *Viscum Caryopylloïdes Maximum flore tripetalo pallide luteo, semine filamentoso*, which is commonly called in that island Wild Pine, Philosoph. Transact. N. 251. Pag. 114. I shall not transcribe the whole description, but only that part of it which relates to this particular. 'From the root (which he had described before) arise leaves on every side, after the manner of leeks or ananas, whence the name of wild pine, or aloes, being folded or enclosed one within another, each of which is two feet and an half long, and from a three inch breadth at beginning or base, ends in a point, having a very hollow or concave inward side, and a round or convex outward one; so that by all their hollow sides is made within a very large preservatory, cistern, or basin, fit to contain a pretty quantity of water, which in the rainy season falls upon the utmost parts of the spreading leaves, which have channels in them, conveying it down to the cistern where it is kept, as in a bottle; the leaves, after they are swelled out like a bulbous root to make the bottle, bending inward or coming again close to the stalk, by that means hindering the evaporation of the water by the heat of the sunbeams.

'In the mountainous as well as the dry low woods, in scarcity of water, this reservatory is not only necessary and sufficient for the nourishment of the plant itself, but likewise is very useful to men, birds and all sorts of insects, whither in scarcity of water they come in troops, and seldom go away without refreshment.

'Captain Dampierre in his voyages, Vol. II. of Campeche, tells us, that these basins made of the leaves of the wild pine will hold a pint and a half, or a quart of water, and that when they find these pines they stick their knives into the leaves just above the roots, and that lets out the water, which they catch in their hats, as (saith he) I have done many times to my great relief.'

Fourthly, The cinnamon-tree of Ceylon, in whose parts there is a wonderful diversity; out of the root they get a sort of camphire, and its oil; out of the bark of the trunk the true oil of cinnamon; from the leaves, an oil like that of cloves; out of the fruit a juniper oil, with a mixture of those of cinnamon and cloves; besides, they boil the berries into a sort of wax, out of which they make candles, plaisters, unguents. Here we may take notice of the candle-trees of the West-Indies, out of whose fruit, boiled to a thick fat consistence, are made very good candles, many of which have been lately distributed by that most ingenious merchant Mr. Charles Dubois.

Fifthly, The fountain or dropping trees, in the Isles of Ferro, St. Thomas, and in Guinea, which serve the inhabitants instead of rain, and fresh springs: my honoured friend Dr. Tancred Robinson, in a late letter to me, is not of Vossius's opinion, that these trees are of the ferulaceous kind, because he observes that by the descriptions of eye-witnesses, and by the dried sample sent by Paludanus to the duke of Wirtemberg, the leaves are quite different from those of the Ferula's, coming nearer to the *Seseli Ethiopicum Salicia vel Percilymeni folio:* therefore the doctor rather thinks them to be of the laurel kind, though he concludes there may be many different sorts of these running aqueous trees, because that phœnomenon does not depend upon, or proceed from any peculiarity of the plant, but rather from the place and situation; of which he writes more at large in a letter printed in another discourse of mine.

Sixthly, and lastly, We will only mention the names of some other vegetables, which, with eighteen or twenty thousand more of that kind, do manifest to mankind the illustrious bounty and providence of the Almighty and Omniscient Creator, towards his undeserving creatures; as the cotton trees; the manyoc, or cassava; the potatoe; the jesuit's bark-tree; the poppy; the rhubarb; the scammony; the jalap; the coloquintida; the china; sarsa; the serpentaria virginiana, or snake-weed, the nisi, or genseg; the numerose balsam, and gum trees, many of which are of late much illustrated by the great industry and skill of that most discerning botanist Dr. Leonard Plukenet. Of what great use all these, and innumerable other plants, are to mankind in the several parts of life, few or none can be ignorant; besides the known uses in curing diseases, in feeding and clothing the poor, in building, in dying, in all mechanics, there may be as many more not yet discovered, and which may be reserved on purpose to exercise the faculties bestowed on man, to find out what is necessary, convenient, pleasant, or profitable to him.

To sum up all in brief: this terraqueous globe, we know, is made up of two parts:

1. A thin and fluid;
2. A firm and consistent.

The former called by the name of water; the latter, of earth, or dry land. The land being the more dense and heavy body, doth naturally descend beneath the water, and occupy the lower place; the water ascends and floats above it. But we see that it is not thus: for the land, though the more heavy, is forcibly, and contrary to its nature, so elevated as to cast off the water, and stand above it, being (as the Psalmist phrases it) founded upon, or ' above the seas, and established above the floods,' Psal. xxiv. 2. And this in such manner, that not only on one side of the globe, but on all sides, there were probably continents and islands raised so equally as to counterbalance one another, the water flowing between them, and filling the hollow and depressed places. Neither was the dry land only raised up, and made to appear, but some parts, which we call mountains, were highly elevated above others; and those so disposed and situated (as we have shewn) in the midland parts, and in continued chains running east and west, as to render all the earth habitable, a great part whereof otherwise would not have been so; but the torrid zone must indeed have been such a place as the ancients fancied it, unhabitable for heat. Let us now consider how much better it is, that the dry land should be thus raised up, and the globe divided almost equally between earth and water, than that all its surface should be one uniform and dark body of water. I say water, because that naturally occupies the superior place, and not earth. For were it all water, the whole beauty of this inferior world were gone : there could be no such pleasant and delicious prospects as the earth now affords us; no distinction and grateful variety of mountains and hills, plains and valleys, rivers, and pools, and fountains : no shady woods stored with lofty and towering trees for timber; lowly, and more spread ones, for shade and fruit : no amicable verdure of herbs, bespangled with an infinite variety of specious and fragrant flowers : for those plants that grow at the bottom of the sea, are, for the most part, of a dull, sullen, and dirty olive colour, and bear no flowers at all. Instead of the elegant shapes and colours, the sagacity and docility of ingenious beasts and birds, the musical voices and accents of the aerial choristers, there had been nothing but mute stupid and indocile fishes, which seem to want the very sense of discipline, as may be gathered from that they are not vocal, and that there appear in them no organs of hearing : it being also doubtful, whether the element they live in be capable of transmiting sounds; the best sense they have, even their sight, can be but dull and imperfect, the element of water being semi-opaque, and reflecting a good part of the beams of light. The most noble and ingenious creatures that live there, the cetaceous kind, being near akin to terrestrial animals, and breathing in the same element, the open air. Had, I say, all been water, there had been no place for such a creature as man ; as we see there is no such there. There is no business for him, no subject to employ his art and faculties, and consequently there could be no effects of them : no such things as houses and cities, and stately edifices ; as gardens and orchards, and walks, and labyrinths, as corn-fields, and vineyards, and the rest of these ornaments, wherewith the wit and industry of man hath embellished the world.

These are great things, and worthy the care and providence of the Creator; which whoso considereth, and doth not discern and acknowledge, must needs be as stupid as the earth he goes upon.

But because mountains have been looked upon by some as warts and superfluous excrescencies, of no use or benefit; nay, rather as signs and proofs, that the present earth is nothing else but a heap of rubbish and ruins, I shall deduce and demonstrate in particulars, the great use, benefit, and necessity of them.

1. They are of eminent use for the production and original of springs and rivers. Without hills and mountains there could be no such things, or at least but very few: no more than we now find in plain and level countries; that is, so few, that it was never my hap to see one. In wintertime, indeed, we might have torrents and landfloods, and perhaps sometimes great inundations, but in summer nothing but stagnating water, reserved in pools and cisterns, or drawn up out of deep wells; but as for a great part of the earth (all lying within, or near the tropics) it would neither have rivers, nor any rain at all. We should consequently lose all those conveniencies and advantages that rivers afford us, of fishing, navigation, carriage, driving of mills, engines, and many others. This end of mountains I find assigned by Mr. Edmund Halley, a man of great sagacity and deep insight into the natures and causes of things, in a discourse of his published in the Philosoph. Transactions, numb. 192. in these words: 'This, if we may al-

low final causes [hardiment, the thing is clear, pronounce boldly without any *ifs* or *ands*] this seems to be one design of the hills, that their ridges being placed through the midst of their continents, might serve as it were alembics, to distil fresh water for the use of man and beast; and their heights to give a descent to those streams, to run gently like so many veins of the macrocosm, to be the more beneficial to the creation.'

2 They are of great use for the generation, and convenient digging up of metals and minerals: which how necessary instruments they are of culture and civility I have before shewn. These we see are all digged out of mountains, and I doubt whether there is, or can be, any generation of them in perfectly plain and level courtries. But if there be, yet could not such mines, without great pains and charges, if at all, be wrought; the delfs would be so flown with waters (it being impossible to make any addits or soughs to drain them) that no gins or machines could suffice to lay and keep them dry.

3 They are useful to mankind in affording them convenient places for habitation, and situations of houses and villages, serving as screens to keep off the cold and nipping blasts of the northern and easterly winds, and reflecting the benign and cherishing sun-beams, and so rendering their habitations both more comfortable, and more cheerly in winter; and promoting the growth of herbs and fruit-trees, and the maturation of their fruits in summer. Besides casting off the waters, they lay the gardens, yards, and avenues to the houses dry and clean; and so as well more salutary, as more elegant. Whereas houses built in plains, unless shaded with trees, lie bleak and exposed to wind and weather:

and all winter are apt to be grievously annoyed with mire and dirt.

4. They are very ornamental to the earth, affording pleasant and delightful prospects, both, 1. To them that look downwards from them, upon the subjacent countries; as they must needs acknowledge, who have been but on the downs of Sussex, and enjoyed that ravishing prospect of the sea on one hand, and the country far and wide on the other. And 2. To those that look upwards and behold them from the plains and low grounds; which what a refreshing and pleasure it is to the eye, they are best able to judge who have lived in the isle of Ely, or other level countries, extending on all sides farther than one can ken; or have been out far at sea, where they can see nothing but sky and water. That the mountains are pleasant objects to behold, appears, in that the very images of them, their draughts and landscapes, are so much esteemed.

5. They serve for the production of great variety of herbs and trees. For it is a true observation, that mountains do especially abound with different species of vegetables, because of the great diversity of soils that are found there, every vertex, or eminency, almost affording new kinds. Now these plants serve partly for the food and sustenance of such animals as are proper to the mountains, partly for medicinal uses; the chief physic herbs and roots, and the best in their kinds, growing there: it being remarkable, that the greatest and most luxuriant species in most genera of plants are native of the mountains: partly also for the exercise and diversion of such ingenious and industrious persons, as are delighted in searching out these natural rarities; and observing the outward form, growth, natures,

and uses of each species, and reflecting upon the Creator of them his due praises and benedictions.

6. They serve for the harbour, entertainment, and maintenance of various animals, birds, beasts, and insects, that breed, feed, and frequent there. For the highest tops and pikes of the alps themselves are not destitute of their inhabitants, the ibex, or stein-buck, the rupicapra, or chamois, among quadrupeds; the lagopus among birds; and I myself have observed beautiful papilio's, and store of other insects, upon the tops of some of the Alpine mountains. Nay, the highest ridges of many of those mountains, serve for the maintenance of cattle for the service of the inhabitants of the valleys : the men there, leaving their wives and younger children below, do, not without some difficulty, clamber up the acclivities, dragging their kin with them, where they feed them, and milk them, and make butter and cheese, and do all the dairy-work, in such sorry hovels and sheds as they build there to inhabit in during the summer months. This I myself have seen and observed in mount Jura, not far from Geneva, which is high enough to retain snow all the winter.

The same they do also in the Grisons country, which is one of the highest parts of the Alps, travelling through which I did not set foot off snow for four days' journey, at the latter end of March.

7. Those long ridges and chains of lofty and topping mountains, which run through whole continents east and west (as I have elsewhere observed), serve to stop the evagation of the vapours to the north and south in hot countries, condensing them, like alembic heads, into water, and so by a kind of external distillation giving original to springs and rivers; and likewise by

amassing, cooling, and constipating of them, turn them into rain; by those means rendering the fervid regions of the torrid zone habitable.

This discourse concerning the use of mountains, I have made use of in another treatise;* but because it is proper to this place, I have with some alterations and enlargements here repeated it.

I had almost forgotten that use they are of to mankind, in serving for boundaries and defences to the territories of kingdoms and commonwealths.

A second particular I have made choice of, more exactly to survey and consider, is the body of man: wherein I shall endeavour to discover something of the wisdom and goodness of God, first, By making some general observations concerning the body. Secondly, By running over and discoursing upon its principal parts and members.

1. Then in general I say, the wisdom and goodness of God appears in the erect posture of the body of man, which is a privilege and advantage given to man above other animals. But though this be so, yet I would not have you think, that all the particulars I shall mention are proper only to the body of man, divers of them agreeing to many other creatures. It is not my business to consider only the prerogatives of man above other animals, but the endowments and perfections which nature hath conferred on his body, though common to them with him. Of this erection of the body of man, the ancients have taken notice, as a particular gift and favour of God.

Pronaque cum spectent animalia cætera terram,
Os homini sublime dedit, cœlumque tueri
Jussit, et erectos ad sidera tollere vultus.—*Ovid. Metam. I.*

---

* The Dissolution of the World.

And before him, Tully in his second book De Nat. Deorum.

'Ad hanc providentiam naturæ tam diligentem tamque solertema djungi multa possunt, e quibus intelligatur quantæ res hominibus a Deo quamque eximiæ tributæ sunt, qui primum eos humo excitatos, celsos, et erectos constituit, ut Deorum cognitionem cœlum intuentes capere possent. Sunt enim e terra homines, non ut incolæ atque habitatores, sed quaei spectatores superarum rerum atque cœlestium, quarum spectaculum ad nullum aliud genus animantium pertinet.'

Man being the only creature in this sublunary world made to contemplate heaven, it was convenient that he should have such a figure or situs of the parts of his body, that might conveniently look upward. But to say the truth in this respect of contemplating the heavens or looking upwards, I do not see what advantage a man hath by this erection above other animals, the faces of most of them being more supine than ours, which are only perpendicular to the horizon, whereas some of theirs stand reclining. But yet two or three other advantages we have of this erection, which I shall here mention.

First, It is more commodious for the sustaining of the head, which, being full of brains and very heavy (the brain in man being far larger in proportion to the bulk of his body, than in any other animal), would have been very painful and wearisome to carry, if the neck had lain parallel or inclining to the horizon.

Secondly, This figure is most convenient for prospect and looking about one. A man may see farther before him, which is no small advantage for avoiding dangers, and discovering whatever he searches after.

Thirdly, The conveniency of this site of our

bodies will more clearly appear, if we consider what a pitiful condition we had been in, if we had been constantly necessitated to stand and walk upon all four; man being by the make of his body, of all quadrupeds (for now I must compare him with them) the most unfit for that kind of incessus, as I shall shew anon. And besides that we should have wanted, at least in a great measure, the use of our hand, that invaluable instrument, without which we had wanted most of those advantages we enjoy as reasonable creatures, as I shall more particularly demonstrate afterward.

But it may be perchance objected by some, that nature did not intend this erection of the body, but that it is superinduced and artificial; for that children at first creep on all four, according to that of the poet,

Mox quadrupes, ritusque tulit sua membra ferarum.—*Ovid.*

To which I answer, that there is so great an inequality in the length of our legs and arms, as would make it extremely inconvenient, if not impossible, for us to walk upon all four, and set us almost upon our heads; and therefore we see that children do not creep upon their hands and feet, but upon their hands and knees; so that it is plain that nature intended us to walk as we do, and not upon all four.

2. I argue from the situs or position of our faces; for had we been to walk upon all four we had been the most prone of all animals, our faces being parallel to the horizon and looking directly downwards.

3. The greatness and strength of the muscles of the thighs and legs above those of the arms, is a clear indication that they were by nature intended for a more difficult and laborious action,

even the moving and transferring the whole body, and that motion to be sometimes continued for a great while together.

As for that argument taken from the contrary flexure of the joints of our arms and legs to that of quadrupeds; as that our knees bend forward, whereas the same joint of their hind legs bends backward, and that our arms bend backward whereas the knees of their fore legs bend forward. Although the observation be as old as Aristotle, because I think there is a mistake in it, in not comparing the same joints (for the first or uppermost joint in a quadruped's hind legs bend forward as well as a man's knees, which answer to it, being the uppermost joint of our legs; and the like, *mutatis mutandis*, may be said of the arms), I shall not insist upon it.

Another particular which may serve to demonstrate that this erect posture of the body of man was intended and designed by the wise and good Author of nature, is the fastening of the cone of the pericardium to the midriff: an account whereof I shall give the reader out of the ingenious Dr. Tyson's 'Anatomy of the Orang-outang, or Pygmy,' p. 49.

Vesalius (saith he) and others make it a peculiarity to man, that the pericardium, or bag that incloses the heart, should be fastened to the diaphragm. Vesalius tells us, (De Corporis humain fabrica, lib. 6. cap 8.) Cæterum involucri murco et dextri ipsius lateris egregia portio septi transversi nerveo circulo validissime amploque admodum spatio connascitur, quod hominibus est peculiare.' The point of the pericardium, and a very considerable portion of its right side, is most firmly fastened to the nervous circle of the midriff for a large space, which is peculiar to mankind. So Blancardius, Anat. reformat.

cap. 2. p. 8. 'Homo præ cæteris animalibus hoc peculiare habet, quod ejus pericardium septi tranversi medio semper accrescat, cum idem in quadrupedum genere liberum et aliquanto spatio ab ipso remotum sit.' 'Man hath this peculiar to him, and different from other animals; that his pericardium doth always grow to the middle of the midriff, whereas in the quadruped kind it is free and removed some distance from it.'

The pericardium in man is therefore thus fastened, that in expiration it might assist the diastole of the diaphragm; for otherwise the liver and stomach being so weighty, they would draw it down too much towards the abdomen, so that, upon the relaxation of its fibres in its diastole, it would not ascend sufficiently into the thorax, so as to cause a subsidency of the lungs by lessening the cavity there. In quadrupeds there is no need of this adhesion of the pericardium to the diaphragm; for in them in expiration, when the fibres of the diaphragm are relaxed, the weight of the viscera of the abdomen will easily press the diaphragm up into the cavity of the thorax, and so perform that service. Besides, were the pericardium fastened to the diaphragm in quadrupeds, it would hinder its systole in inspiration, or its descent downwards upon the contraction of its muscular fibres, and the more, because the diaphragm being thus tied up, it could not then so freely force down the weight of the viscera, which are always pressing upon it, and consequently not sufficiently dilate the cavity of the thorax, and therefore must hinder their inspiration. Thus we see how necessary it is, that in man, the pericardium should be fastened to the diaphragm, and in quadrupeds how inconvenient it would be. And since we find this difference between the hearts of brutes and men in this particular, how can we imagine but that it must needs be the effect of wisdom and design, and that man was intended by nature to walk erect, and not upon all four as quadrupeds do?

II. The body of man may thence be proved to be the effect of wisdom, because there is nothing in it deficient, nothing superfluous, nothing but hath its end and use. So true are those maxims we have already made use of, 'Natura nihil facit frustra;' and 'Natura non abundat in superfluis nec deficit in necessariis,' no part that we we can well spare. 'The eye cannot say to the hand I have no need of thee, nor the head to the feet I have no need of you;' 1 Cor. xii. 21. that I may usurp the apostle's similitude.

The belly cannot quarrel with the members, nor they with the belly for her seeming sloth; as they provide meat for her, so she concocts and distributes it to them. Only it may be doubted to what use the paps in men should serve. I answer, partly for ornament, partly for a kind of conformity between the sexes, and partly to defend and cherish the heart; in some they contain milk, as in a Danish family we read of in Bartholine's Anatomical Observations. However, it follows not that they or any other parts of the body are useless, because we are ignorant.

I have lately met with a story in Seignior Paulo Boccone's Natural Observations, printed at Bologna, in Italy, 1684. well attested, concerning a countryman called Billardino di Billo, living in a village belonging to the city of Nocera in Umbria, called Somareggio, whose wife dying, and leaving a young infant, he nourished it with his own milk. This man, either because in the small village where he lived there was not a wet nurse to be had, or because he was not able to hire one, took the child, and applying it to his own bosom, and putting the nipples of his breasts into its mouth, invited it to suck, which the infant did, and after divers time drawing, fetched some milk. Whereat the father encouraged, continued to apply it, and so after a while it brought down the milk so plentifully as to nourish it for many months, till it was fit to be weaned. Hereupon my author having proved by sufficient authority of able anatomists, such as Franciscus Maria Florentinus, and Marcellus Malpighius, that the paps of men have the same structure and the same vessels with those of women, concludes, that nature hath not given paps to men, either to no purpose, or for mere ornament, but if need requires, to supply the defect of the female, and give suck to the young.

Had we been born with a large wen upon our faces, or a Bavarian poke under our chins, or a great bunch upon our backs like camels, or any the like superfluous excrescency, which should be not only useless but troublesome, not only stand us in no stead, but also be ill-favoured to behold, and burthensome to carry about, then we might have had some pretence to doubt whether an intelligent and bountiful Creator had been our architect; for had the body been made by chance, it must in all likelihood have had many of these superfluous and unnecessary parts.

But now, seeing there is none of our members but hath its place and use, none that we could spare, or conveniently live without, were it but those we account excrements, the hair of our heads, or the nails on our fingers' ends; we must needs be mad or sottish, if we can conceive any other than that an infinitely good and wise God was our author and former.

III. We may fetch an argument of the wisdom and providence of God from the convenient situation and disposition of the parts and members of our bodies: they are seated most conveniently for use, for ornament, and for mutual assistance. First, for use; so we see the senses of such eminent use for our well-being, situate in the head, as sentinels in a watch-tower, to receive and convey to the soul the impressions of external objects. 'Sensus autem interpretes ac nuntii rerum in capite, tanquam in arce, mirifice ad usus necessarios et facti et collocati sunt.' Cic. de Nat. Deorum. The eye can more easily see things at a distance; the ear receives sounds from afar. How could the eye have been better placed either for beauty and ornament, or for the guidance and direction of the whole body? As Cicero proceeds well, ' Nam oculi tanquam speculatores altissimum locum obtinent, ex quo plurima conspicientes funguntur suo munere: et aures quæ sonum recipere debent, qui natura in sublime fertur, recte in altis corporum partibus collocatæ sunt; itemque nares, eo quod omnis odor ad superiora fertur, recte sursum sunt.' 'For the eyes like sentinels occupy the highest place, from whence seeing many things they perform their functions: and the ears, which are made for the reception of sounds, which naturally are carried upwards, are rightly placed in the uppermost parts of the body; also the nostrils, because all odours ascend, are fitly situate in the superior parts.' I might instance in the other members. How could the hands have been more conveniently placed for all sorts of exercises and works, and for the guard and security of the head and principal parts? The heart to dispense life and heat to the whole body, viz. near the centre; and yet because it is harder for the blood to ascend than descend, somewhat nearer the head. It is also observable, that the sinks of the body are removed as far from the nose and eyes as may be; which Cicero takes notice of in the fore-mentioned place. 'Ut in ædificiis architecti avertunt ab oculis et naribus dominorum ea quæ profluentia necessario essent tetri aliquid habitura, sic natura res similes procul amandavit a sensibus.'

Secondly, for ornament. What could have been better contrived, than that those members which are pairs, should stand by one another in equal altitude, and answer on each side one to another? and,

Thirdly, for mutual assistance. We have before shewed how the eye stands most conveniently for guiding the hand, and the hand for defending the eye; and the like might be said of the other parts, they are so situate as to afford direction and help one to another. This will appear more clearly if we imagine any of the members situate in contrary places or positions. Had a man's arms been fitted only to bend backwards behind him, or his legs only to move backwards; what direction could his eyes then have afforded him in working or walking? Or how could he then have fed himself? Nay, had one arm been made to bend forward, and the other directly backward, we had then lost half the use of them, since they could not have assisted one another in any action. Take the eyes, or any other of the organs of sense, and see if you can find any so convenient a seat for them in the whole body as that they now possess.

IV. From the ample provision that is made for the defence and security of the principal parts: Those are, 1. The heart; which is the fountain of life and vegetation: 'Officina spirituum vitalium, principium et fons caloris nativi, lucerna humidi radicalis,' and that I may speak with the chymists, 'ipse sol microcosmi,' the very *sun* of the microcosm, or little world, in which is contained that vital flame or heavenly fire, which Prometheus is fabled to have stole from Jupiter; or, as Aristotle phrases it, that 'Ανάλογον τῷ τῶν ἀπλανῶν στοιχείῳ, 'Divinum quid respondens elemento stellarum.' This, for more security, is situate in the centre of the trunk of the body, covered first with its own membrane, called pericardium, lodged within the soft bed of the lungs, encompassed round with a double fence, (1.) Of firm bones or ribs to bear off blows: (2.) Of thick muscles and skins, besides the arms conveniently placed to fence off any violence at a distance, before it can approach to hurt it.

2. The brain, which is the principle of all sense and motion, the fountain of the animal spirits, the chief seat and palace royal of the soul; upon whose security depends whatever privilege belongs to us as sensitive or rational creatures. This, I say, being the prime and immediate organ of the soul, from the right constitution whereof proceeds the quickness of apprehension, acuteness of wit, solidity of judgment, method and order of invention, strength and power of memory; which if once weakened and disordered, there follows nothing but confusion and disturbance in our apprehensions, thoughts, and judgments, is environed round about with such a potent defence, that it must be a mighty force indeed that is able to injure it.

1. A skull so hard, thick, and tough, that it is almost as easy to split a helmet of iron as to make a fracture in it. 2. This covered with skin and hair, which serve to keep it warm, being naturally a very cold part, and also to quench and dissipate the force of any stroke that shall be dealt it, and retund the edge of any weapon. 3. And yet more than all this, there is still a thick and tough membrane which hangs looser

about it, and doth not so closely embrace it (that they call *dura mater*), and in case the skull happens to be broken doth often preserve it from injury and diminution: and lastly, a thin and fine membrane strait and closely adhering to keep it from quashing and shaking. The many pairs of nerves proceeding from it, and afterward distributing and branching themselves to all the parts of the body either for nutrition or motion, are wonderful to behold in prepared bodies, and even in the schemes and figures of Dr. Willis and Vieussen.

I might instance, (3.) in the lungs, which are so useful to us as to life and sense, that the vulgar think our breath is our very life, and that we breathe out our souls from thence. Suitable to which notion, both *anima* and *spiritus* in Latin, and πνεῦμα in Greek, are derived from words that signify breath and wind: and *efflare*, or *exhalare animam*, signify, to die. And the old Romans used to apply mouth to mouth, and receive the last gasps of their dying friends, as if their souls had come out that way. From hence perhaps might first spring that opinion of the vehicles of spirits; the vulgar, as I hinted before, conceiving that the breath was, if not the soul itself, yet that wherein it was wafted and carried away. These lungs, I say, are for their better security and defence shut up in the same cavity with the heart.

V. In the abundant provision that is made against evil accidents and inconveniences. And the liberality of nature appears, 1. In that she hath given many members, which are of eminent use, by pairs, as two eyes, two ears, two nostrils, two hands, two feet, two breasts [*mammæ*], two reins: that so, if by any cross or unhappy accident one should be disabled or rendered useless, the other might serve us tolerably well; whereas had a man but one hand, or one eye, &c. if that were gone, all were gone, and we left in evil case. See then and acknowledge the benignity of the Deity, who hath bestowed upon us two hands, and two eyes, and other the like parts, not only for our necessity, but conveniency, so long as we enjoy them; and for our security, in case any mischance deprive us of one of them. 2. In that all the vessels of the body have many ramifications: which particular branches, though they serve mainly for one member or muscle, yet send forth some twigs to the neighbouring muscles; and so interchangeably the branches that serve these, send to them. So that if one branch chance to be cut off or obstructed, its defect may in some measure be supplied by the twigs that come from the neighbouring vessels. 3. In that she hath provided so many ways to evacuate what might be hurtful to us, or breed diseases in our bodies. If any thing oppress the head, it hath a power to free itself by sneezing: if any thing fall into the lungs, or if any humour be discharged upon them, they have a faculty of clearing themselves and casting it up by coughing: if any thing clog or burden the stomach, it hath an ability of contracting itself and throwing it up by vomit. Besides these ways of evacuation, there are siege, urine, sweating, hæmorrhages from the nose and hæmorrhoidal veins, fluxes of rheum. Now the reason why nature hath provided so many ways of evacuation, is because of the different humours that are to be voided or cast out. When therefore there is a secretion made of any noxious humour, it is carried off by that emunctory whose pores are fitted to receive and transmit the minute parts of it; if at least this separation be made by percolation, as we will now suppose, but not assert. Yet I doubt not but the same humour may be cast off by divers emunctories, as is clear in urine and sweat, which are for the main the same humour carried off several ways.

To this head of provision against inconveniences, I shall add an observation or two concerning sleep.

1. Sleep being necessary to man and other animals for their refreshment, and for the reparation of that great expense of spirits which is made in the day-time, by the constant exercise of the senses and motions of the muscles, that it might ease and refresh us indeed, nature hath provided that, though we lie long upon one side, we should have no sense of pain or uneasiness during our rest, no nor when we awake. Whereas in reason one would think, that the whole weight of the body pressing the muscles and bones on which we lie, should be very burthensome and uneasy, and create a grievous sense of pain, and we find by experience that it doth so when we lie long awake in the night, we being not able (especially if never so little indisposed) to rest one quarter of an hour in the same posture without shifting of sides, or at least etching this way and that way, more or less. How this may be effected is a great question. To me it seems most probable, that it is done by an inflation of the muscles whereby they become both soft, and yet renitent like so many pillows, dissipating the force of the pressure, and so preventing or taking away the sense of pain. That the muscles are in this manner inflated in time of rest, appears to the very eye in the faces of children, and may be proved from that when we rest in our clothes we are fain to loosen our garters, shoe-strings, and other ligatures to give the spirits free pas-

sage, else we shall experience pain and uneasiness in those parts, which when we are waking we find not.

The reason of this phenomenon, viz. that ἀναλγησία, or want of pain we experience in sleep, during and after a long *decubitus* on one side, Dr. Liter in his 'Journey to Paris,' p. 113. and Dr. Jones in his 'Treatise of the Mysteries of Opium revealed,' attribute to the relaxation of the nerves and muscles in time of sleep; and the sense of pain and uneasiness when we lie awake to the tension of them during that time. This I do not deny, but yet I think the reason I have assigned hath a great interest in that rest and easiness we enjoy when asleep.

2. Because sleep is inconsistent with the sense of pain, therefore during rest, those nerves which convey that motion to the brain, which excites in the soul a sense of pain, are obstructed. This I myself have had frequent experience of since I have been troubled with sores on my legs: upon sudden awakening, finding myself at perfect ease, and void of all sense of pain for a minute's time or more, the pain then by degrees returning, which I could attribute to nothing else but the dissipating that vapour, or whatever else it were, which obstructed the nerves, and giving the dolorific motion free passage again.

Upon second thoughts, and reading what Dr. Lister and Dr. Jones have written concerning this subject, I rather incline to believe that the motion causing a sense of pain is conveyed to the brain by the nerves themselves in tension, as we see in chords, any the least touch at one end passes speedily to the other when they are stretched, which it doth not when they are relaxed, and not by the spirits passing through them: and on the other side, the unsensibleness of pain proceeds rather from the relaxation of the nerves than their obstruction. But yet this tension of the nerves and muscles is owing to the spirits flowing down into them and distending them.

VI. From the constancy that is observed in the number, figure, place, and make of all the principal parts; and from the variety in the less. Man is always mending and altering his works; but nature observes the same tenor, because her works are so perfect, that there is no place for amendments; nothing that can be reprehended. The most sagacious men in so many ages have not been able to find any flaw in these divinely contrived and formed machines, 'no blot or error in this great volume of the world, as if any thing had been an imperfect essay at the first,' to use the Bishop of Chester's words: nothing that can be altered for the better; nothing but if it were altered would be marred. This could not have been, had man's body been the work of chance, and not counsel and providence. Why should there be constantly the same parts? Why should they retain constantly the same places? Why should they be endued with the same shape and figure? Nothing so contrary as constancy and chance. Should I see a man throw the same number a thousand times together upon but three dice, could you persuade me that this were accidental, and that there was no necessary cause of it? How much more incredible then is it, that constancy in such a variety, such a multiplicity of parts, should be the result of chance? Neither yet can these works be the effects of necessity or fate, for then there would be the same constancy observed in the smaller as well as the larger parts and vessels; whereas there we see nature doth *ludere,* as it were, sport itself; the minute ramifications of all the vessels, veins, arteries, and nerves, infinitely varying in individuals of the same species, so that they are not in any two alike.

VII. The great wisdom of the divine Creator appears, in that there is pleasure annexed to those actions that are necessary for the support and preservation of the *individuum*, and the continuation and propagation of the species; and not only so, but pain to the neglect or forbearance of them. For the support of the person, it hath annexed pleasure to eating and drinking; which, else out of laziness, or multiplicity of business, a man would be apt to neglect, or sometimes forget. Indeed, to be obliged to chew and swallow meat daily for two hours' space, and to find no relish or pleasure in it, would be one of the most burthensome and ungrateful tasks of a man's whole life. But because this action is absolutely necessary, for abundant security nature hath inserted in us a painful sense of hunger to put us in mind of it, and to reward our performance, hath adjoined pleasure to it. And as for the continuation of kind, I need not tell you, that the enjoyments which attend those actions are the highest gratifications of sense.

VIII. The wonderful art and providence of the Contriver and Former of our bodies, appears in the multitude of intentions he must have in the formation of the several parts, or the qualifications they require, to fit them for their several uses. Galen,[*] in his book *De formatione fœtus*, takes notice, that there are in a human body above six hundred several muscles, and there are at least ten several intentions or due qualifications to be observed in each of these; proper figure, just magnitude, right disposition of its several ends,

[*] Bishop of Chester's Nat. Rel. lib. i. c. 6.

upper and lower, position of the whole, the insertion of its proper nerves, veins, and arteries, which are each of them to be duly placed; so that about the muscles alone no less than six thousand several ends or aims are to be attended to. The bones are reckoned to be 284. The distinct scopes or intentions in each of these are above 40, in all about 100,000. And thus it is in some proportion with all the other parts, the skin, ligaments, vessels, glandules, humours: but more especially with the several members of the body, which do, in regard of the great variety and multitude of those several intentions required to them, very much exceed the homogeneous parts. And the failing in any one of these would cause irregularity in the body, and in many of them such as would be very notorious.' Now to imagine that such a machine, composed of so many parts, to the right form, order, and motion whereof such an infinite number of intentions are required, could be made without the contrivance of some wise agent, must needs be irrational in the highest degree.

This wonderful mechanism of human bodies, next to viewing the life, may be seen at large in the excellent figures of Spigelius and Bidloo; their situation, order, connexion, and manner of separating them in Lyserus's Cult. Anatom. The almost infinite ramifications and inosculations of all the several sorts of vessels, the structures of the glands, and other organs, may easily be detected by glasses, and traced by blowing in of air and drying them, or by injecting through peculiar syringes, melted wax, or quicksilver; the operations whereof may be learnt out of Swammerdam, Caspar Bartholine, and Antonio Nuck.

IX. Another argument of wisdom and design

in the contrivance of the body of man and other animals, is the fitting of some parts to divers offices and uses, whereby nature doth (as the proverb is) ' Una fidelia duos parietes dealbare ;' ' Stop two gaps with one bush.' So, for instance, the tongue serves not only for tasting, but also to assist the mastication of the meat and deglutition, by turning it about and managing it in the mouth; to gather up the food in man by licking; in the dog and cat-kind by lapping; in kine by plucking up the grass; particularly in man, it is of admirable use for the formation of words and speaking.

The diaphragm and muscles of the abdomen or lower belly, are of use not only for respiration, but also for compressing the intestines, and forcing the chyle into the lacteal veins, and likewise out of the said veins into the thoracic channel. And here, to note that by the way, appears the use of a common receptacle of chyle, that by the motion of the muscles of respiration, it being pressed upon, the chyle might with more facility be impelled into the forementioned duct. Besides, this action of respiration, and motion of the said diaphragm and muscles, may serve also for the comminution and concoction of the meat in the stomach (as some not without reason think) by their constant agitation and motion upwards and downwards, resembling the pounding or braying of materials in a mortar.

And to instance in no more, the muscular contraction and pulse of the heart serves not only for the circulation of the blood, but also for the more perfect mixture of its parts, preserving its due crasis and fluidity, and incorporating the chyle and other juices it receives with it.

X. The wisdom and goodness too of the divine Former of our bodies appears in the nourishment of them. For that food which is of a wholesome juice, and proper to nourish and preserve them in a healthful state, is both pleasant to the taste, and grateful and agreeable to the stomach, and continues to be so till our hunger and thirst be well appeased, and then begins to be less pleasant, and at last even nauseous and loathsome. 'The full stomach loathes the honeycomb.'

On the other side, that which is unwholesome and unfit for nourishment, or destructive of health, is also unpleasant to the taste and ungrateful and disagreeable to the stomach, and that more or less according as it is more or less improper or noxious. And though there be some sorts of food less pleasant to the taste, which by use may be rendered grateful, yet to persons that are in health, and in no necessity of using such viands, I think it were better to abstain from them, and follow nature in eating such things as are agreeable to their palate and stomach. For such unpleasant diet must needs alter the temper of the body before it can become acceptable; and doubtless for the worse.

I might add hereto, that even pain, which is the most grievous and afflictive thing that we are sensible of, is of great use to us. God hath annexed a sense of pain to all diseases and harms of the body inward and outward (and there is no pain but proceeds from some harm or disease), to be an effectual spur to excite and quicken us to seek for speedy help or remedy, and hath so ordered it, that as the disease heals, so the pain abates. Neither doth pain provoke us only to seek ease and relief when we labour under it, but also makes us careful to avoid for the future all such things as are productive of it, that is, such things as are hurtful to our bodies and destructive of the health and well-being of them, which also are for the most part prohibited by God, and so sinful and injurious to our souls. So we see what cure the divine Providence hath taken, and what effectual means it hath used for the healing of our diseases, and the maintenance and preservation of our health. This is the true reason of pain. Howbeit, I will not deny but that God doth sometimes himself immediately inflict diseases, even upon his own children, for many good considerations which I shall not here enumerate. Neither shall I mention the uses that parents and masters make of it for the correcting their children and servants, or magistrates for the punishing of malefactors, they being beyond my scope, only I cannot but take notice, that it is a πολύχρηστον, a thing of manifold uses, and necessary for the government both of commonwealths and families.

XI. Some fetch an argument of providence from the variety of lineaments in the faces of men, which is such, that there are not two faces in the world absolutely alike; which is somewhat strange, since all the parts are in specie the same. Were nature a blind architect, I see not but the faces of some men might be as like, as eggs laid by the same hen, or bullets cast in the same mould, or drops of water out of the same bucket. This particular I find taken notice of by Pliny in his seventh book, cap. 1. in these words: 'Jam in facie vultuque nostro cum sint decem aut paulo plura membra, nullas duas in tot millibus hominum indiscretas effigies exsistere, quod ars nulla in paucis numero præstet affectando;' to which, among other things, he thus prefaces, 'Naturæ vero rerum vis atque majestas in omnibus momentis fide caret.'

Though this at first may seem to be a matter

of small moment, yet, if duly considered, it will appear to be of mighty importance in all human affairs; for should there be an undiscernable similitude between divers men, what confusion and disturbance would necessarily follow? what uncertainty in all sales and conveyances, in all bargains and contracts? what frauds and cheats, and suborning of witnesses? what a subversion of all trade and commerce? what hazard in all judicial proceedings? in all assaults and batteries, in all murders and assassinations, in thefts and robberies, what security would there be to malefactors? Who could swear that such and such were the persons that committed the facts, though they saw them never so clearly? Many other inconveniences might be instanced in; so that we see this is no contemptible argument of the wisdom and goodness of God.

Neither is the difference of voices less considerable for the distinguishing of sexes and particular persons, and individuals of all animals, than that of faces; as Dr. Cockburn makes out, Essay, &c. part ii. p. 68, &c. Nay, in some cases more; for hereby persons in the dark, and those that are blind, may know and distinguish one another, which is of great importance to them; for otherwise they might be most grossly cheated and abused.

Farther we may add out of the same author, p. 71. 'And to no other cause (than the wise providence of God) can be referred the no less strange diversity of hand-writings. Common experience shews, that though hundreds and thousands were taught by one master, and one and the same form of writing, yet they should all write differently. Whether men write court or Roman hand, or any other, there is something peculiar in every one's writing which distinguisheth it, Some indeed can counterfeit another's character and subscription, but the instances are rare, nor is it done without pains and trouble. Nay, the most expert and skilful cannot write much so exactly like, as that it cannot be known whether it be genuine or counterfeit. And if the providence of God did not so order it, what cheats and forgeries too would daily be committed, which would not only justle private men out of their rights, but also unhinge states and governments, and run all into confusion? The diversity of hand-writing is of mighty great use to the peace of the world: it prevents fraud, and secures men's property; it obligeth the living and present, to honesty and faithfulness; it importeth the mind of the absent, and sheweth the will of the dead, which ought to be sacredly observed. And what is so very useful is not the effect of any human concert: men did not of themselves agree to it, they are only carried to it by the secret providence of God, who understandeth and mindeth what is for the good and interest of mankind in general, and of every particular person.'

Add farther to all this, That whereas there are several parts peculiar to brutes, which are wanting in man; as for example, the seventh or suspensory muscle ef the eye, the nictating membrane, the strong aponeuroses on the sides of the neck, called by some *packwax*, it is very remarkable, that these parts are of eminent and constant use to them, as I shall particularly shew hereafter, but to man would have been altogether useless and superfluous.

I have done with my general observations. I proceed now more accurately and minutely to consider some particular parts or members of the body; and first, The head, because it was to contain a large brain, made of the most capacious figure, as near as could be to a spherical; upon this grows the hair, which though it be esteemed an excrement, is of great use (as I shewed before) to cherish and keep warm the brain, and to quench the force of any stroke that might otherwise endanger the skull. It serves also to disburden the brain of a great deal of superfluous moisture wherewith it abounds.

I find it remarked by Marchetti, a famous anatomist in Padua, that the cause of baldness in men is the dryness of the brain, and its shrinking from the 'cranium' or skull; he having observed that in bald persons, under the bald part, there was always a vacuity or empty space between the skull and the brain. And lastly, to name no more, it serves also for a graceful ornament to the face, which our present age is sensible enough of, bestowing so much money upon false hair and periwigs.

Secondly, Another member which I shall more particularly treat of, is the eye, a part so artificially composed, and commodiously situate, as nothing can be contrived better for use, ornament, or security, nothing to advantage added thereto or altered therein. Of the beauty of the eye I shall say little, leaving that to poets and orators; that it is a very pleasant and lovely object to behold, if we consider the figure, colours, and splendour of it, is the least that I can say. The soul, as it is more immediately and strongly moved and affected by this part than any other, so doth it manifest all its passions and perturbations by this. As the eyes are the windows to let in the species of all exterior objects into the dark cells of the brain, for the information of the soul; so are they flaming torches to reveal to those abroad, how the soul within is moved or affected. These representations made by the impressions of external objects upon the eye are the most clear, lively, and distinct of any others. Now to this use and purpose of informing us what is abroad round about us in this aspectable world, we shall find this structure and mechanism of the eye, and every part thereof, so well fitted and adapted, as not the least curiosity can be added. For first of all, the humours and tunicles are purely transparent, to let in the light and colours unfoiled and unsophisticated by any inward tincture. It is usually said by the peripatetics, that the crystalline humour of the eye (which they ineptly fancied to be the immediate organ of vision, wherein all the species of external objects were terminated) is without all colour, because its office was to discern all colours, or at least, to receive the species of several colours, and convey them to the common sense. Now if itself had been coloured, it would have transmitted all visible objects tinctured with the same colour; as we see whatever is beheld through a coloured glass, appears of the same colour with the glass, and to those that have the jaundice or the like suffusion of eyes, objects appear of that same colour wherewith their eyes are infected. This they say is in a great measure true, although they are much mistaken about the organ and manner of vision, and the uses of the humours and membranes of the eye. Two reasons therefore may be assigned why all the membranes and humours of the eye are perfectly pellucid and void of colour; first, For the clearness: secondly, For the distinctness of vision.

First, The clearness. For had the tunicles and humours of the eye, all or any of them, been colorate, many of the rays proceeding from the visible object would have been stopped and suffocated before they could come to the bottom of the eye, where the formal organ of vision is situate. For it is a most certain rule, how much any body hath of colour, so much hath it of opacity, and by so much the more unfit is it to transmit the species.

Secondly, For the distinctness of vision. For, as I said before, and the peripatetics observe well, were the humours of the eye tinctured with any colour, they would refund that colour upon the object, and so it would not be represented to the soul, as in itself it is. So we see that through a coloured glass things appear as well more dim and obscure, as tinctured with the colour thereof.

Secondly, The parts of the eye are made convex, and especially the crystalline humour, which is of a lenticular figure, convex on both sides; that by the refractions there made, there might be a direction of many rays coming from one point in the object, viz. as many as the pupil can receive, to one point answerable in the bottom of the eye; without which the sense would be very obscure, and also confused. There would be as much difference in the clearness and distinction of vision, were the outward surface of the tunica cornea plain, and the crystalline humour removed; as between the picture received on a white paper in a dark room through an open or empty hole, and the same received through a hole furnished with an exactly polished lenticular crystal; which, how great it is, any one that hath but seen this experiment made, knows well enough. Indeed this experiment doth very much explain the manner of vision; the hole answering to the pupil of the eye, the crystalline humour to the lenticular glass, the dark room to the cavity containing the vitreous humour, and the white paper to the tunica retina.

Thirdly, The uveous coat or iris of the eye hath a musculous power, and can dilate and contract that round hole in it, called the pupil, or sight of the eye. It contracts it for the excluding superfluous light, and preserving the eye from being injured by too vehement and lucid an object, and again dilates it for the apprehending objects more remote, or placed in a fainter light; ' tam miro artificio,' saith Scheiner, 'quam munifica naturæ largitate.' If any one desires to make experiment of these particulars, he may, following Scheiner's and Des Cartes's directions, take a child, and setting a candle before him, bid him look upon it; and he shall observe his pupil to contract itself very much, to exclude the light, with the brightness whereof it would otherwise be dazzled and offended; as we are when, after we have been some time in the dark, a bright light is suddenly brought in and set before us, till the pupils of our eyes have gradually contracted themselves. Let the candle be withdrawn, or removed aside, he shall observe the child's pupil by degrees to dilate itself. Or let him take a bead or the like object, and holding it near the eye, command the child to look at it, the pupil will contract much when the object is near; but let it be withdrawn to a greater distance in the same light, and he shall observe the pupil to be much enlarged.

Fourthly, The uveous coat, and also the inside of the choroïdes, are blackened like the walls of a tennis-court, that the rays may be there suffocated and suppressed, and not reflected backwards to confound the sight. And if any be by the retiform coat reflected, they are soon choked in the black inside of the uvea. Whereas were they reflected to and fro, there could be no distinct vision. As we see the light admitted into the dark room we even now spake of, obliterates the species which before were seen upon the white cloth or paper.

Fifthly, Because the rays from a nearer and from a more remote object do not meet just in the same distance behind the crystalline humour (as may easily be observed in lenticular glasses, where the point of concourse of the rays from a nearer object is at a greater distance behind the glass, and from a farther at a lesser), therefore the ciliary processes, or rather the ligaments observed in the inside of the sclerotic tunicles of the eye by a late ingenious anatomist, do serve instead of a muscle, by their contraction, to alter the figure of the eye, and make it broader, and consequently draw the retina nearer to the crystalline humour, and by their relaxation suffer it to return to its natural distance according to the exigency of the object, in respect of distance or propinquity. And besides possibly the ciliary processes may by their constriction or relaxation, render the crystalline itself more gibbose or plain; and with the help of the muscles a little alter the figure of the whole eye for the same reason. To what I have said might be added, that the retiform tunicle is whitish, for the better and more true reception of the species of things. That there being a distance necessarily required for the collection of the rays received by the pupil, viz. those that proceed from one point of the object to one point again in the bottom of the eye, the retine must needs be set at a distance from the crystalline humour; and therefore nature hath provided a large room, and filled it with the pellucid vitreous humour most fit for that purpose.

I must not omit a notable observation concerning the place of the insertion of the optic nerve into the bulb of the eye, and the reason of it; which I owe to that learned mathematician Peter Herigon: 'Nervus opticus,' saith he, in his Optica, 'ad latus ponitur, ne pars imaginis in ejus foramen indicens pictura careat.' The optic nerve is not situate directly behind the eye, but on one side, lest that part of the image that falls upon the hole of the optic nerve, should want its picture. This I do not conceive to be the true reason of this situation; for even now as it is situate, that part of the object whose rays fall upon the centre or hole of the optic nerve, wants its picture, as we find by experience; that part not being seen by us, though we heed it not. But the reason is, because if the optic axis should fall upon this centre (as it would do, were the nerve seated just behind the eye) this great inconvenience would follow, that the middle point of every object we viewed would be invisible, or there would a dark spot appear in the midst of it. Thus we see the admirable wisdom of nature in thus placing the optic nerve in respect of the eye; which he that did not consider or understand would be apt to think more inconveniently situate for vision, than if it had been right behind.

Another thing also concerning vision is most remarkable, that though there be a decussation of the rays in the pupil of the eye, and so the image of the object in the retina or bottom of the eye be inverted, yet doth not the object appear inverted, but in its right or natural posture. The reason whereof is, because the visual rays coming in straight lines, by those points of the sensory or retina which they touch, affect the common sense or soul, according to their direction; that is, signify to it that those several parts of the object from whence they proceed lie in straight lines, point for point, drawn through the pupil to the several points of the sensory where they terminate, and which they press upon. Whereupon the soul must needs conceive the object, not in an inverted, but a right posture. And that the nerves are naturally made not only to inform the soul of external objects which press upon them, but also of the situation of such objects, is clear, because if the eyes be distorted, the objects, will we, nill we, will apear double. So if the fore and middle fingers be crossed, and a round body put between them and moved, it will seem to be two; the reason is, because in that posture of the fingers the body touches the outsides of them, which in their natural site are distant one from another, and their nerves made to signify to the soul bodies separate and distant in like manner, two fingers lying between them. And though our reason by the help of our sight corrects this error, yet cannot we but fancy it to be so.

Neither is the aqueous humour, as some may supinely imagine, altogether useless or unprofitable as to vision, because by its help the uvea tunica is sustained, which else would fall flat upon the crystalline humour; and fluid it must be, to give way to the contraction and dilatation of the uveous: and because the outermost coat of the eye might chance to be wounded, or pricked, and this humour, being fluid, let out, therefore nature hath made provision speedily to repair it again in such a case, by the help of certain water-pipes, or lymphæ-ducts inserted into the bulb of the eye, proceeding from glandules designed by nature to separate this water from the blood for that use. Antonius Nuck affirms, that if the eye of an animal be pricked, and the aqueous humour squeezed out, in ten hours' space the said humour and sight shall be restored to the eye, if at least the creature be kept in a dark place: and that he did publicly demonstrate the same in the anatomical theatre at Leyden, in a dog, out of whose eye, being wounded, the aqueous humour did so copiously flow, that the membranes appeared flaccid, and yet in six hours' space the bulb of the eye was again replete with its humour, and that without the application of any medicines. Antonius Nuck 'de Ductu novo salivali,' &c.

Moreover, it is remarkable, that the cornea tunica (horny or pellucid coat of the eye) doth not lie in the same superficies with the white of the eye, but riseth up, as it were a hillock, above its convexity, and is of an hyperbolical or parabolical figure. So that though the eye seems to be perfectly round, in reality it is not so, but the iris thereof is protuberant above the white; and the reason is, because that if the cornea tunica, or crystalline humour, had been concentrical to the sclerodes, the eye could not have admitted a whole hemisphere at one view, 'et sic animalis incolumitati in multis rebus minus cautum esset,' as Scheiner well observes. In many things there had not been sufficient caution or care taken for the animal's safety.

And now (that I may use the words of a late author[*] of our own) the eye is already so perfect, that I believe the reason of a man would easily have rested here, and admired at his own contrivance. For he being able to move his whole body upward and downward, and on every side, might have unawares thought himself sufficiently well provided for; but nature hath add-

[*] Dr. Moore's Antidote against Atheism.

ed muscles also to the eyes, that no perfection might be wanting; for we have often occasion to move our eyes, our head being unmoved, as in reading and viewing more particularly an object set before us, by transferring the axes of our eyes all over it. And that this may be done with the more ease and accuracy, she hath furnished this organ with no less than six muscles, to move it upward, and downward, to the right and left, obliquely and round about.

I shall now consider what provision is made for the defence and security of this most excellent and useful part.

First, The eyes are sunk in a convenient valley, 'latent utiliter,' and are encompassed round with eminent parts, as with a rampart, ' et excelsis undique partibus sepiuntur;' Cic.* so are defended from the strokes of any flat or broad bodies. Above stand the eyebrows to keep off any thing from running down upon them, as drops of sweat from the forehead, or dust, or the like. 'Superiora superciliis obducta sudorem a capite et fronte defluentem repellunt.' Cic. Then follow the eyelids, which fence them from any sudden and lesser stripes. These also round the edges are fortified with stiff bristles, as it were palisadoes, against the incursions of importunate animals, serving partly as a fan to strike away flies or gnats, or any other troublesome insect ; and partily to keep off superfluous light. 'Munitæque sunt palpebræ tanquam vallo pilorum, quibus et apertis oculis siquid incideret repelleretur.' Idem, Ibid. And because it was necessary that man and other animals should sleep, which could not be so well done if the light came in by the windows of the eyes, therefore hath nature provided these curtains to

* De Natur. Deor. l. 2.

be then drawn to keep it out. And because the outward coat of the eye ought to be pellucid to transmit the light, which if the eyes should always stand open, would be apt to grow dry and shrink, and lose their diaphaneity, therefore are the eye-lids so contrived as often to wink, that so they may as it were glaze and varnish them over with the moisture they contain, there being glandules on purpose to separate a humour for that use, and withal wipe off whatever dust or filth may stick to them; and this, lest they should hinder the sight, they do with the greatest celerity. Cicero hath taken notice that they are made very soft, lest they should hurt the sight: 'Mollissimæ tactu ne læderent aciem, aptissime factæ et ad claudendas pupillas ne quid incideret, et ad aperiendas, idque providit ut identidem fieri posset maxima cum celeritate.'

Secondly, If we consider the bulb or ball of the eye, the exterior membrane or coat thereof is made thick, tough, and strong, that it is a very hard matter to make a rupture in it; and besides so slippery, that it eludes the force of any stroke, to which also its globular figure gives it a very great advantage.

Lastly, Because for the guidance and direction of the body in walking and any exercise, it is necessary the eye should be uncovered, and exposed to the air at all times and in all weathers, therefore the most wise Author of nature hath provided for it a hot bed of fat which fills up the interstices of the muscles; and besides made it more patient and less sensible of cold than our other parts; and though I cannot say with Cicero, absolutely free from danger or harm by that enemy, yet least obnoxious to the injuries thereof of any part, and not at all unless it be immoderate and extreme.

To all this I might add the convenience of the situation of the eye in respect of its proximity to the brain, the seat of apprehension and common sense. Whereas had it been farther removed, the optic nerves had been liable to many more dangers and inconveniences than now they are.

Seeing then the eye is composed of so great variety of parts all conspiring to the use of vision, whereof some are absolutely necessary, others very useful and convenient, none idle or superfluous, and which is remarkable, many of them of a different figure and consistency from any others in the body besides, as being transparent, which it was absolutely necessary they should be, to transmit the rays of light; who can but believe that this organ was designed and made purposely for the use for which it serves?

Neither is it to be esteemed any defect or imperfection in the eyes of man that they want the seventh muscle, or the nictating membrane, which the eyes of many other animals are furnished withal; for though they be very useful, and in a manner necessary to them, considering their manner of living, yet they are not so to man. To such beasts as feed upon grass and other herbs, and therefore are forced to hold their eyes long in a hanging posture, and to look downwards for the choosing and gathering of their food, the seventh or suspensory muscle is very useful, to enable them to do so without much pain or weariness; yet to man, who doth not, nor hath any occasion, indeed cannot hold his head or look long downwards, it would be useless and superfluous. As for the nictating membrane, or periophthalmium, which all birds, and I think most quadrupeds, are furnished with, I have been long in doubt what the use of it might be; and have sometimes thought it was for the more abundant defence and security of the eye; but then I was puzzled to give any tolerable account why nature should be more solicitous for the preservation of the eyes of brutes than men, and in this respect also to be a step-mother to the most noble creature.

But the honourable author* formerly mentioned, gives a probable account why frogs and birds are furnished with such a membrane. Frogs, because being amphibious animals, designed to pass their lives in watery places, which for the most part abound with sedges, and other plants endowed with sharp edges or points; and the progressive motion of this animal being to be made not by walking, but by leaping, if his eyes were not provided of such a sheath, he must either shut them, and so leap blindly and by consequence dangerously, or by leaving them open run a venture to have the cornea cut, pricked, or otherwise offended by the edges or points of the plants, or what may fall from them upon the animal's eye; whereas this membrane (being something transparent as well as strong) is like a kind of spectacle that covers the eye without taking away the sight. Birds are likewise furnished with it, because being destinated to fly among the branches of trees and bushes, their prickles, twigs, leaves, or other parts, would be apt otherwise to wound or offend their eyes. But yet still we are to seek why it is given to other quadrupeds, whose eyes are in no such danger.

Thirdly, The ear, another organ of sense, how admirably is it contrived for the receiving and conveying of sounds? First, There is the outward ear or auricula, made hollow and con-

* Boyle of Fin. Causes, p. 53, 54.

tracted by degrees to draw the sound inward, to take in as much as may be of it, as we use a funnel to pour liquor into any vessel. And therefore if the auricula be cut clear off, the hearing is much impaired, and almost quite marred, as hath been by experience found. From the auricula is extended a small, long, round hole inward into the head, to intend the motion and so augment the force of the sound, as we see in a shooting trunk, the longer it is to a certain limit, the swifter and more forcibly the air passes in it, and drives the pellet. At the end of this hole is a membrane, fastened to a round bony limb, and stretched like the head of a drum, and therefore by anatomists called also tympanum, to receive the impulse of the sound, and to vibrate or quaver according to its reciprocal motions or vibrations; the small ear-bones being at the end fastened to the tympanum, and furnished with a muscle, serve for the tension of that membrane, or the relaxation of it according to the exigency of the animal, it being stretched to the utmost when it would hearken diligently to a lower or more distant sound. Behind the drum are several vaults and anfractuose cavities in the earbone, filled only with what naturalists call the implanted air; so to intend the least sound imaginable, that the sense might be affected with it; as we see in subterraneous caves and vaults how the sound is redoubled, and what a great report it makes however moderate it be. And because it was for the behoof of the animal, that upon any sudden accident it might be awakened out of its sleep, therefore were there no shuts or stopples made for the ears, that so any loud or sharp noise might awaken it, as also a soft and gentle voice of murmur provoke it to sleep. Now the ears for the benefit and conveniences of the animal being always to stand open, because there was some danger that insects might creep in thereat, and eating their way through the tympanum, harbour in the cavities behind it; therefore hath nature loricated or plaistered over the sides of the forementioned hole with ear-wax, to stop and entangle any insects that should attempt to creep in there. But I must confess myself not sufficiently to understand the nature of sounds to give a full and satisfactory account of the structure and uses of all the parts of the ear. They who have a mind to search into the curious anatomy and use of this part, may consult Monsieur Du Verney.

Fourthly, The next part I shall take notice of shall be the teeth, concerning which I find seven observations in the honourable Mr. Boyle's Treatise of Final Causes, which I shall briefly recapitulate, and add one or two more.

1. That the teeth alone among the bones continue to grow in length during a man's whole life, as appears by the unsightly length of one tooth when its opposite happens to fall or be pulled out; which was most providently designed to repair the waste that is daily made of them by the frequent attrition in mastication. Here by the bye I might advise men to be careful how they attempt to cure this blemish by filing or cutting off the head of such an overgrown tooth, lest that befal them which happened to a certain nun in Padua, who, upon cutting off a tooth in that manner, was presently convulsed and fell into an epilepsy, as Bartholine in his anatomy reports.

2. That that part of the teeth which is extant above the gums is naked, and not invested with that sensible membrane called periosteum, wherewith the other bones are covered.

3. That the teeth are of a closer and harder substance than the rest of the bones, for the more easy breaking and comminution of the more solid aliments, and that they might be more durable, and not so soon worn down by grinding the food.

4. That for the nourishing and cherishing these so necessary bones, the all-wise Author of things has admirably contrived an unseen cavity in each side of the jaw-bone, in which greater channel are lodged an artery, a vein, and a nerve, which through lesser cavities, as it were through gutters, send their twigs to each particular tooth.

5. Because infants were for a considerable time to feed upon milk, which needs no chewing, and lest teeth should hurt the tender nipples of the nurse, nature hath deferred the production of them for many months in a human fœtus; whereas those of divers other animals, which are reduced to seek betimes food that needs mastication, are born with them.

6. The different figure and shape of the teeth is remarkable. That the fore-teeth should be formed broad and with a thin and sharp edge like chizzles, to cut off and take away a morsel from any solid food, called therefore 'incisores.' The next, one on each side stronger and deeper rooted, and more pointed, called therefore 'canini,' in English eye-teeth, to tear the more tough and resisting sort of aliments. The rest called jaw-teeth or grinders, in Latin 'molares,' are made flat and broad at top and withal somewhat uneven and rugged, that by their knobs and little cavities they may the better retain, grind, and commix the aliments.

7. Because the operations to be performed by the teeth oftentimes require a considerable

firmness and strength, partly in the teeth themselves, partly in the instruments which move the lower jaw, which alone is moveable, nature hath provided this with strong muscles, to make it bear forcibly against the upper jaw; and thus not only placed each tooth in a distinct cavity of the jaw-bone, as it were in a close, strong, and deep socket, but has furnished the several sorts of teeth with hold-fasts suitable to the stress that by reason of their different offices they are to be put to. And therefore, whereas the cutters and eye-teeth have usually but one root (which in these last named is wont to be very long), the grinders that are employed to crack nuts, stones of fruit, bones, or other hard bodies, are furnished with three roots, and in the upper jaw often with four, because these are pendulous, and the substance of the jaw somewhat softer.

8. The situation of the teeth is most convenient, viz. the molares or grinders behind, nearest the centre of motion, because there is a greater strength or force required to chew the meat, than to bite a piece; and the cutters before, that they may be ready to cut off a morsel from any solid food, to be transmitted to the grinders.

9. It is remarkable that the jaw in men and such animals as are furnished with grinders, hath an oblique or transverse motion, which is necessary for chewing and comminution of the meat; which it is observed not to have in those animals that want the molares.

Now if, as Galen saith, he that shall marshal a company but of thirty-two men in due order, is commended for a skilful and industrious person, shall we not admire nature which hath so skilfully ranked and disposed this quire of our teeth?

Fifthly, The tongue is no less admirable for the contexture and manifold uses of it. First, It is the organ of tasting; for being of a spongy substance, the small particles of our meat and drink being mingled with the saliva, easily insinuate themselves into the pores of it, and so do either gratefully affect it, or harshly grate upon it, accordingly as they are figured and moved; and hereby we discern what is convenient or inconvenient for our nourishment. It helps us likewise in the chewing and swallowing of our meat: and lastly, It is the main instrument of speaking, a quality so peculiar to man, that no beast could ever attain to it. And although birds have been taught to form some words, yet they have been but a few, and those learn with great difficulty; but what is the chief, the birds understand not the meaning of them, nor use them as signs of things or their own conceptions of them; though they may use them as expressions of their passions; as parrots having been used to be fed at the prolation of certain words, may afterward when they are hungry pronounce the same. For this Des Cartes makes his main argument to prove that brutes have no cogitation, because the highest of them could never be brought to signify their thoughts or conceptions by any artificial signs, either words, or gestures (which, if they had any, they would in all likelihood be forward enough to do), whereas all men, both fools and mutes, make use of words or other signs to express their thoughts, about any subjects that present themselves; which signs also have no reference to any of their passions. Whereas the signs that brute animals may be taught to use are no other than such as are the motions of some of their passions, fear, hope, joy, &c.

Hence some of the Jewish rabbins did not so absurdly define a man* 'animal loquens,' a speaking creature. Having had occasion just now to mention the saliva or spittle, I am put in mind of the eminent use of this humour, which is commonly taken for an excrement. Because a great part of our food is dry, therefore nature hath provided several glandules to separate this juice from the blood, and no less than four pair of channels to convey it into the mouth, which are of late invention, and called by anatomists 'ductus salivales,' through which the saliva distilling continually, serves well to macerate and temper our meat, and make it fit to be chewed and swallowed. If a copious moisture did not by these conduit-pipes incessantly flow down into the mouths of horses and kine, how were it possible they should for a long time together grind and swallow such dry meat as hay and straw? Moreover it is useful not only in the mouth but in the stomach too, to promote concoction, as we have already noted.

Sixthly, To the mouth succeeds the wind-pipe, no less wonderful in its conformation. For because continual respiration is necessary for the support of our lives, it is made with annulary cartilages to keep it constantly open, and that the sides of it may not flag and fall together. And lest when we swallow, our meat or drink should fall in there and obstruct it, it hath a strong shut or valve called epiglottis, to cover it close, and stop it when we swallow. For the more convenient bending of our necks, it is not made of one entire continued cartilage, but of many annular ones joined together by strong

* חי כוכת.

membranes, which membranes are muscular, compounded of strait and circular fibres, for the more effectual contraction of the wind-pipe in any strong or violent expiration or coughing. And lest the asperity or hardness of these cartilages should hurt the œsophagus or gullet, which is tender and of a skinny substance, or hinder the swallowing of our meat, therefore these annulary gristles are not made round, or entire circles; but where the gullet touches the wind-pipe, there to fill up the circle is only a soft membrane, which may easily give way to the dilatation of the gullet. And to demonstrate that this was designedly done for this end and use, so soon as the wind-pipe enters the lungs, its cartilages are no longer deficient, but perfect circles or rings, because there was no necessity they should be so, but it was more convenient they should be entire. Lastly, for the various modulation of the voice, the upper end of the wind-pipe is endued with several cartilages and muscles, to contract or dilate it as we would have our voice flat or sharp; and moreover the whole is continually moistened with a glutinous humour issuing out of the small glandules that are upon its inner coat, to fence it against the sharp air received in, or breath forced out; yet it is of quick and tender sense, that it may be easily provoked to cast out by coughing whatever may fall into it from without, or be discharged into it from within.

It is also very remarkable which Caspar Bartholine hath observed in the gullet, that where it perforateth the midriff, the carneous fibres of that muscular part are inflected and arcuate, as it were a sphincter embracing and closing it fast, by a great providence of nature, lest in the perpetual motion of the said midriff, the upper orifice of the stomach should gape, and cast out the victuals as fast as it received it.

Seventhly, The heart which hath been always esteemed, and really is, one of the principal parts of the body, the 'primum vivens, et ultimum moriens,' the first part that quickens and the last that dies, by its incessant motion distributing the blood, the vehicle of life, and with it the vital heat and spirits, throughout the whole body, whereby it doth continually irrigate, nourish, and keep hot and supple all the members. Is it not admirable that from this fountain of life and heat there should be channels and conduit-pipes, to every, even the least and most remote, part of the body? just as if from one water-house there should be pipes conveying the water to every house in a town, and to every room in each house; or from one fountain in a garden there should be little channels or dikes cut to every bed, and every plant growing therein, as we have seen more than once done beyond the seas. I confess the heart seems not to be designed to so noble a use as is generally believed, that is, to be the fountain or conservatory of the vital flame, and to inspire the blood therewith (for the lungs serve rather for the accension or maintaining that flame, the blood receiving there from the air those particles which are one part of the pabulum or fuel thereof, and so impregnated running back to the heart); but to serve as a machine to receive the blood from the veins, and to force it out by the arteries through the whole body, as a syringe doth any liquor, though not by the same artifice. And yet this is no ignoble use, the continuance of the circulation of the blood being indispensably necessary for the quickening and enlivening of all the members of the body, and supplying of matter to the brain, for the preparation of the animal spirits, the instruments of all sense and motion. Now for this use of receiving and pumping out of the blood, the heart is admirably contrived. For, first, being a muscular part, the sides of it are composed of two orders of fibres running circularly or spirally from base to tip, contrarily one to the other, and so being drawn or contracted contrary ways, do violently constringe and straiten the ventricles, and strongly force out the blood, as we have formerly intimated. Then the vessels we call arteries, which carry from the heart to the several parts, have valves which open outwards like trap-doors, and give the blood a free passage out of the heart, but will not suffer it to return back again thither, and the veins, which bring it back from the several members to the heart, have valves and trap-doors which open inwards, so as to give way unto the blood to run into the heart, but prevent it from running back again that way. Besides, the arteries consist of a quadruple coat, the third of which is made up of annular or orbicular carneous fibres to a good thickness, and is of a muscular nature, after every pulse of the heart, serving to contract the vessel successively with incredible celerity, so by a kind of peristaltic motion impelling the blood onwards to the capillary extremities, and through the muscles, with great force and swiftness. So the pulse of the arteries is not only caused by the pulsation of the heart, driving the blood through them in manner of a wave or flush, as Des Cartes and others would have it; but by the coats of the arteries themselves, which the experiments of a certain Lovain physician* (the first whereof is Galen's), do in my

* Cartes, Ep. v. 1. Ep. 77. et seq.

opinion make good against him. First, saith he, if you slit the artery and thrust into it a pipe, so big as to fill the cavity of it, and cast a strait ligature upon that part of the artery containing the pipe, and so bind it fast to the pipe; notwithstanding the blood hath free passage through the pipe, yet will not the artery beat below the ligature; but do but take off the ligature, it will commence again to beat immediately. But because one might be ready to reply to this experiment, that the reason why when bound it did not beat, was because the current of the blood, being straitened by the pipe, when beneath the pipe it came to have more liberty, was not sufficient to stretch the coat of the artery, and so cause a pulse, but when the ligature was taken off, it might flow between the enclosed tube and the coat of the artery; therefore he adds another, which clearly evinces that this could not be the reason, but that it is something flowing down the coats of the artery that causes the pulse: that is, if you straighten the artery never so much, provided the sides of it do not quite meet, and stop all passage of the blood, the vessel will notwithstanding continue still to beat below or beyond the coarctation. So we see some physicians both ancient (as Galen) and modern, were of opinion, that the pulse of the arteries was owing to their coats; though the first that I know of who observed the third coat of an artery to be a muscular body, composed of annular fibres, was Dr. Willis. This mention of the peristaltic motion puts me in mind of an ocular demonstration of it in the gullet of kine when they chew the cud, which I have often beheld with pleasure. For after they have swallowed one morsel, if you look steadfastly upon their throat, you will soon see another ascend, and run pretty swiftly all along the throat up to the mouth, which it could not do unless it were impelled by the successive contraction or peristaltic motion of the gullet, continually following it. And it is remarkable that these ruminant creatures have a power by the imperium of their wills of directing this peristaltic motion upwards or downwards. I shall add no more concerning the heart, but that it and the brain do 'mutuas operas tradere,' enable one another to work; for first the brain cannot itself live, unless it receive continual supplies of blood from the heart, much less can it perform its functions of preparing and distributing the animal spirits; nor the heart pulse, unless it receive spirits or something else that descends from the brain by the nerves. For do but cut asunder the nerves that go from the brain to the heart, the motion thereof in more perfect and hot creatures ceaseth immediately. Which part began this round is the question.

I find in the Philosophical Transactions, No. 280. some notable observations of the famous anatomist Mr. William Cowper, concerning the artifice of nature in regulating the motion of the blood in the veins and arteries to assist and promote it in the one, and moderate it in the other, which I shall give you in his own words.

As the arteries (saith he) are known to export the blood, so the veins to carry it back again to the heart; but having already described their extremities, we come now to the large trunks of the veins, and here, as in the arteries, we find the common practice of nature in disposing the branches of veins to discharge the refluent blood into the next adjacent trunk, and so on to the heart. As the arteries afford abundance of instances of checks given to the velocity of the current of blood through several parts, so the veins supply us with as many artifices to assist its regular return to the heart, as well as favour those contrivances in the arteries.

The carotid, vertebral, and splenic arteries are not only variously contorted, but also here and there dilated to moderate the motion of the blood, so the veins that correspond to those arteries are also variously dilated. The beginnings of the internal jugulars have a bulbous cavity, which are diverticula to the refluent blood in the sinus's of the dura mater, lest it should descend too fast into the jugulars. The like has been taken notice of by Dr. Lower in the vertebral sinus's. The splenic vein has divers cells opening into it near its extremities in human bodies, but in quadrupeds the cells open into the trunks of the splenic veins.

The spermatic veins do more than equal the length of the arteries of the testes in men; their various divisions and several inosculations, and their valves, are admirably contrived to suspend the weight of the blood, in order to discharge it into the larger trunks of the veins; and were it not that the refluent blood from the testes is a pondus to the influent blood from the arteries, and still lessens its current in the testes, these spermatic veins, like those those of other parts, might have discharged the blood into the next adjacent trunk.

Who can avoid surprise at the art of nature in contriving the veins that bring part of the refluent blood from the lower parts of the body, when they consider the necessity of placing the human heart, as well as that of most quadrupeds, so far from the centre of the body towards its upper part? It is for that end necessary that the large trunks of the veins and arteries should not associate each other. For if all the blood sent to the lower parts by the descending trunk of the aorta should return to the heart again by one single trunk (as it is sent out from thence), the weight of so much blood in the ascending trunk of the vena cava, would oppose all the force the heart could give it from the arteries, and hinder its ascent: for this reason the vena azygos, or *sine pari*, is contrived to convey the blood sent to the muscles of the back and thorax into the descending trunk of the vena cava above the heart. Hence it is evident that more blood comes into the heart by the descending or upper trunk of the vena cava, then passes out by the ascending trunk of the aorta. Nor does the quantity of blood conveyed to the heart by the superior trunk of the cava seem without some other design in nature besides transporting it thither, to free the inferior trunk from its weight. But perhaps it was necessary so much blood should be ready there to join with the chyle for its better mixture before it reaches the right auricle of the heart. So far Mr. Cowper.

Eighthly, the next part I shall treat of shall be the hand, this ὄργανον ὀργάνων, or superlative instrument, which serves us for such a multitude of uses, as it is not easy to enumerate; whereto if we consider the make and structure of it, we shall find it wonderfully adapted. First, it is divided into four fingers bending forward, and one opposite to them bending backwards, and of greater strength than any of them singly, which we call the thumb, to join with them severally or united; whereby it is fitted to lay hold of objects of any size or quantity. The least things, as any small single seed, are taken up by the thumb and forefinger; those a little greater, by the thumb and two fingers, which also we chiefly employ to manage the needle in sowing, and the pen in writing: when we would take up a greater quantity of any thing, we make use of the thumb and all the fingers. Sometimes we use one finger only, as in pointing at any thing, picking things out of holes or long and narrow vessels; sometimes all severally at one time, as in stopping the strings when we play upon any musical instruments. 2. The fingers are strengthened with several bones, jointed together for motion, and furnished with several muscles and tendons, like so many pullies, to bend them circularly forwards; which is most convenient for the firm holding and griping of any object: which of how great, constant, and necessary use it is in pulling or drawing, but especially in taking up and retaining any sort of tool or instrument to work withal in husbandry and all mechanic arts, is so obvious to every man's observation, that I need not spend time to instance in particulars. Moreover, the several fingers are furnished with several muscles to extend and open the hand, and to move them to the right and left: and so this division and motion of the fingers doth not hinder but that the whole hand may be employed, as if it were all of a piece: as we see it is, either expanded, as in striking out, smoothing and folding up of clothes, and some mechanic uses; or contracted, as in fighting, kneading of dough, and the like. It is also notable, and indeed wonderful, that the tendons, bending the middle joint of the fingers, should be perforated to give passage to the tendons of the muscles which draw the uppermost joints, and all bound down close to the bone with strong fillets, lest they should start up and hinder the hand in its work, standing like so many bow strings. 3. The fingers' ends are strengthened with nails, as we fortify the ends of our staves or forks with iron hoops or ferules, which nails serve not only for defence but for ornament, and many uses. The skin upon our fingers ends is thin and of most exquisite sense, to help us judge of any thing we handle. If now I should go about to reckon up the several uses of this instrument, time would sooner fail me than matter. By the help of this we do all our works, we build ourselves houses to dwell in; we make ourselves garments to wear; we plough and sow our grounds with corn, dress and cultivate our vineyards, gardens, and orchards, gather and lay up our grain and fruits; we prepare and make ready our victuals. Spinning, weaving, painting, carving, engraving, and that divinely invented art of writing, whereby we transmit our own thoughts to posterity, and converse with and participate the observations and inventions of them that are long ago dead, all performed by this. This is the only instrument for all arts whatsoever; no improvement to be made of any experimental knowledge without it. Hence (as Aristotle saith well) they do amiss that complain that man is worse dealt with by nature than other creatures; whereas they have some hair, some shells, some wool, some feathers, some scales, to defend themselves from the injuries of the weather, man alone is born naked and without all covering. Whereas they have natural weapons to defend themselves and offend their enemies, some horns, some hoofs, some teeth, some talons, some claws, some spurs and beaks; man hath none of all these, but is weak, and feeble, and unarmed sent into the world. Why, a hand, with reason to use it, supplies the uses of all these; that is both a horn, and a hoof, and a talon, and a tusk, &c. because it enables us to use weapons of these and other fashions, as swords, and spears, and guns. Besides, this advantage a man hath of them, that whereas they cannot at pleasure change their coverings, or lay aside their weapons, or make use of others as occasion serves, but must abide winter and summer, night and day, with the same clothing on their backs, and sleep with their weapons upon them; a man can alter his clothing according to the exigency of the weather, go warm in winter, and cool in summer, cover up himself hot in the night, and lay aside his clothes in the day, and put on or off more or fewer according as his work and exercise is: and can, as occasion requires, make use of divers sorts of weapons, and choice of such at all turns as are most proper and convenient; whereby we are enabled to subdue and rule over all other creatures, and use for our own behoof those qualities wherein they excel, as the strength of the ox, the valour and swiftness of the horse, the vigilancy of the dog, and so make them as it were our own. Had we wanted this member in our bodies, we must have lived the life of brutes, without house or shelter but what the woods and rocks would have afforded; without clothes or covering; without corn, or wine, or oil, or any other drink but water; without the warmth and comfort, or other uses of fire, and so without any artificial baked, boiled, or roast meats; but must have scrambled with the wild beasts for crabs, and nuts, and acorns, and such other things as the earth puts forth of her own accord. We had lain open and exposed to injuries, and had been unable to resist or defend ourselves against almost the weakest creature.

The remaining parts I shall but briefly run over

That the backbone should be divided into so many vertebres for commodious bending, and not be one entire rigid bone, which being of that length would have been often in danger of snapping in sunder. That it should be tapering in form of a pillar, the lower vertebres being the broadest and largest, and the superior in order, lesser and lesser, for the greater firmness and stability of the trunk of the body. That the several vertebres should be so elegantly and artificially compacted and joined together, that they are as strong and firm, as if they were but one bone. That they should be all perforated in the middle with a large hole for the spinal marrow or pith to pass along; and each particular have a hole on each side to transmit the nerves to the muscles of the body, to convey both sense and motion. That by reason of the fore-mentioned close connexion of the vertebres, it should be so formed, as not to admit any great flexure or recess from a right line, any angular, but only a moderate circular bending; lest the spinal pith should be compressed, and so the free intercourse or passage of the spirits to and fro be stopped.

One observation relating to the motion of the bones in their articulations, I shall here add, that is, the care that is taken, and the provision that is made, for the easy and expedite motion of them; there being to that purpose a twofold liquor prepared for the inunction and lubrification of their heads or ends: 1. An oily one, furnished by the marrow. 2. A mucilaginous, supplied by certain glandules seated in the articulations, both which together make up the most apt and proper mixture for this use and end that can be invented or thought upon. For not only both the ingredients are of a lubricating nature, but there is this advantage gained from their composition, that they do mutually improve one another: for the mucilage adds to the lubricity of the oil, and the oil preserves the mucilage from inspissation, and contracting the consistency of a jelly. Now

IN THE CREATION.

this inunction is useful, indeed necessary, for three ends chiefly.

1. For the facilitating of motion. For though the ends of the bones are very smooth, yet were they dry, they could not with that readiness and ease, nay, not without great difficulty, yield to and obey the plucks and attractions of the motory muscles; as we see clocks and jacks, though the screws and teeth of the wheels and nuts be never so smooth and polished, yet if they be not oiled, will hardly move, though you clog them with never so much weight; but if you apply but a little oil, they presently whirl about very swiftly with the tenth part of the force.

2. For preserving the ends of the bones from an incalescency, which they, being hard and solid bodies, would necessarily contract from a swift and long continuing motion: such as that of running, or mowing, or threshing, or sawing, and the like, if they immediately touched and rubbed against one another with that force they must needs do; especially in running, the whole weight of the body bearing upon the joints of the thighs and knees: so we see in the wheels of waggons or coaches, the hollows of the naves by their swift rotations on the end of the axle-trees produce a heat, sometimes so intense, as to set them on fire, to prevent which they stand in need to be frequently anointed or besmeared with a mixture of grease and tar, imitating the forementioned natural composition of oil and mucilage. Nay, bodies softer a great deal than metals contract a great heat by attrition; as is evident from those black circular lines we see on boxes, dishes, and other turned vessels of wood, which are the effects of ignition, caused by the pressure of an edged stick upon the vessel turned nimbly in the lathe. And if there had not been a pro-

vision in the joints against such a preternatural incalescence upon their violent motion, this would have made a slothful world, and confined us to leisurely and deliberate movements, when there were the most urgent and hasty occasions to quicken us.

3. For the preventing of attrition and wearing down the ends of the bones by their motion and rubbing one against another, which is so violent and lasting sometimes, that it is a wonder any inunction should suffice to secure their heads from wasting and consumption. I have often seen the tops of the teeth (which are of a harder substance than the rest of the bones) worn off by mastication, in persons who have lost most of their grinders, and been compelled constantly to make use of three or four only in chewing, so low, that at last the inward marrow and nerve lay bare, and they could no longer for pain make use of them. So that had there not been this provision made for the anointing the bones, the curious workmanship of nature in adapting them so exactly one to another, as was most fit for the easy performance of all those motions to which they were destined, would not suffice for use: but the stirring part of mankind would soon find themselves fitter for an hospital, than for action and the pursuit of business.

These observations I acknowledge myself to have borrowed of a late ingenious writer* of osteology, who thus concludes his discourse upon this subject. 'And here we cannot avoid the notice of the visible footsteps of an infinite reason, which as they are deeply impressed upon the universe, so more especially on the sensible parts of it in those rational contrivances which are found in animals: and we can never sufficiently

* Mr. Clopton Havers.

admire the wisdom and providence of our great Creator, who has given all parts in these animated beings, not only such a structure as renders them fit for their necessary motions and designed functions, but withal the benefit and advantage of whatever may preserve them, or facilitate their action.

Moreover, the artifice of nature is wonderful in the construction of the bones that are to support the body, and to bear great burdens, or to be employed in strong exercises, they being made hollow, for lightness and stiffness. For, we have before noted, a body that is hollow may be demonstrated to be more rigid and inflexible, than a solid one of the same substance and weight. So that here is provision made both for the stiffness and lightness of the bones. But the ribs, which are not to bear any great weight, or to be strongly exercised, but only to fence the breast, have no cavity in them, and towards the forepart or breast are broad and thin, that so they might bend and give way without danger of fracture; when bent returning by their elastic property to their figure again. Yet is not the hollow of the bones altogether useless, but serves to contain the marrow; which supplies an oil for the maintaining and inunction of the bones and ligaments, and so facilitating their motion in the articulations; and particularly (which we mentioned not before) of the ligaments, preserving them from dryness and rigidity, and keeping them supple and flexible, and ready to comply with all the motions and postures of that moveable part to which they appertain; and lastly, to secure them from disruption, which, as strong as they are, they would be in some danger of, upon a great and sudden stretch or contortion, if they were dry, &c. See more to this purpose in the treatise fore-quoted, p. 183.

That whereas the breast is encompassed with ribs, the belly is left free ; that it might give way to the motion of the midriff in respiration ; and to the necessary reception of meat and drink ; as also for the convenient bending of the body ; and in females for that extraordinary extension that is requisite in the time of their pregnancy.

That the lungs should be made up of such innumerable air-pipes and vesicles interwoven with blood vessels in order to purify, ferment, or supply the sanguinous mass with nitro-aerial particles, which rush in by their elastic power upon the muscular extension of the thorax, and so feed the vital flame and spirits ; for upon obstructing this communication, all is presently extinct, no circulation, no motion, no heat, nor any sign of life remains.

That the stomach should be membranous, and capable of dilatation and contraction, according to the quantity of meat contained in it, that it should be situate under the liver, which by its heat might cherish it, and contribute to concoction : that it should be endued with an acid or glandulous ferment, or some corruptive quality for so speedy a dissolution of the meat, and preparation of chyle! that after concoction it should have an ability of contracting itself and turning out the meat.

That the guts should immediately receive it from the pylorus, further elaborate, prepare, and separate it, driving by their peristaltic motion the chyle into the lacteals, and the excrementitious parts to the podex, from whence there is no regress, unless when the valve of the colon is torn and relaxed ; but for the curious structure of these parts, see more in Kerkringius, Glisson, Willis, and Peyer.

That the bladder should be made of a membranous substance, and so extremely dilatable for receiving and containing the urine, till opportunity of emptying it; that it should have shuts for the ends of the ureters so artificially contrived as to give the urine free entrance, but to stop all passage backward, so that they will not transmit the wind, though it be strongly blown and forced in.

That the liver should continually separate the choler from the blood, and empty it into the intestines, where there is good use for it, not only to provoke dejection, but also to attenuate the chyle and render it so subtile and fluid, as to enter in at the orifices of the lacteous veins.

That in the kidneys there should be such innumerable little siphons or tubes conveying the urinose particles to the pelvis and ureters, first discovered by Bellini, and illustrated by Malpighi; that indeed all the glands of the body should be congeries of various sorts of vessels curled, circumgyrated, and complicated together, whereby they give the blood time to stop and separate through the pores of the capillary vessels into the secretory ones, which afterward all exonerate themselves into one common ductus; as may be seen in the works of Dr. Wharton, Graaf, Bartholine, Rudback, Bilsius, Malpighi, Nuck, and others. That the glands should separate such variety of humours all different in colour, taste, smell, and other qualities.

Finally, that all the bones, and all the muscles, and all the vessels of the body, should be so admirably contrived, and adapted, and compacted together for their several motions and uses, and that most geometrically, according to the strictest rules of mechanics; that if in the whole body you change the figure, situation, and conjunction but of one part, if you diminish or increase the bulk and magnitude; in fine, if you endeavour any innovation or alteration, you mar and spoil instead of mending. How can all these things put together but beget wonder and astonishment?

In the muscles alone there seems to be more geometry, than in all the artificial engines in the world; and therefore the different motions of animals, are a subject fit only for the great mathematicians to handle; amongst whom, Steno, Dr. Croon, and above all Alphonso Borelli, have made their essays towards it.

That under one skin there should be such infinite variety of parts, variously mingled, hard with soft, fluid with fixt, solid with hollow, those in rest with those in motion, some with cavities as mortises to receive, others with tenons to fit those cavities; all these so packed and thrust so close together, that there is no unnecessary vacuity in the whole body, and yet so far from clashing or interfering one with another, or hindering each other's motions, that they do all friendly conspire, all help and assist mutually one the other, all concur in one general end and design, the good and preservation of the whole, are certainly arguments and effects of infinite wisdom and counsel; so that he must needs be worse than mad that can find in his heart to imagine all these to be casual and fortuitous, or not provided and designed by a most wise and intelligent cause.

Every part is clothed, joined together, and corroborated by membranes, which upon several occasions (as extravasations of humours, compressions or obstructions of vessels) are capable of a prodigious extension, as we see in the hydatides of the female testicles or ovaries, in hydropical tumours of the lymphæ-ducts, of the scrotum and peritonæum, out of the last of which alone twenty and even forty gallons of water have been drawn by a paracentesis or tapping, for which we have the undoubted authority of Tulpius, Meekren, Pechlin, Blasius, and other medical writers. What vast sacks and bags are necessary to contain such a collection of water, which seems to issue from the lymphæducts, either dilacerated or obstructed, and exonerating themselves into the foldings, or between the duplicatures of the membranes?

Those parts which one would think were of little use in the body, serving chiefly to fill up empty spaces, as the fat, if examined strictly, will be found very beneficial and serviceable to it. 1. To cherish and keep it warm by hindering the evaporation of the hot steams of blood, as clothes keep us warm in winter, by reflecting and doubling the heat. 2. To nourish and maintain the body for some time when food is wanting, serving as fuel to preserve and continue the natural heat of the blood, which requires an oily or sulphureous pabulum, as well as fire. Hence upon long abstinence and fasting, the body grows lean: hence also some beasts, as the marmotto, or mus alpinus, a creature as big or bigger than a rabbit, which absconds all winter, doth (as Hildanus tells us) live upon its own fat. For in the autumn, when it shuts itself up in its hole (which it digs with its feet like a rabbit, making a nest with hay or straw, to lodge itself warm) it is very fat; [Hildanus took out above a pound and half of fat between the skin and muscles, and a pound out of the abdomen] but on the contrary, in the spring-time when it comes forth again, very lean, as the hunters experience in those they then take. 3. The internal fat serves for the defence and security of the vessels, that they might lie soft, and be safely conveyed in their passage, wherefore it is especially gathered about them.

By what pores, or passages, or vessels, the fat is separated from the blood when it is redundant, and again absorbed into it when it is deficient, is a matter of curious inquiry, and worthy to be industriously sought out by the most sagacious and dexterous anatomists. The vessels whereinto it is received, and wherein contained, are by the microscope detected to be bladders, and those doubtless perforated and pervious one into another; and though for their excessive subtlety and thinness they appear not in a lean body, yet seem to have been primitively formed and provided by nature to receive the fat upon occasion. Why the fat is collected chiefly about some particular parts and vessels, and not others, as for example, the reins and the caul, I easily consent with Galen and others, the reason to be the cherishing and keeping warm of those parts upon which such vessels are spread; so the caul serves for the warming the lower belly, like an apron or piece of woollen cloth. Hence a certain gladiator, whose caul Galen cut out, was so liable to suffer from the cold, that he was constrained to keep his belly constantly covered with wool. For the intestines containing a great deal of food, there to undergo its last concoction, and no vessels of blood penetrating it, and flowing through it to keep it warm, they had need be defended from the injuries of the external air, by outward coverings. Why there should be such copious fat gathered about the reins to enclose them, is not so easy to discern; but surely there is a great and constant heat required there for the separation of the urine from the blood; the constant separation and excretion whereof, is necessary for the preservation of life. And we see if the blood be in any degree chilled, the secretion of urine is in a great measure stopt, and the serum cast upon the glandules of the mouth and throat. And if the blood be extraordinarily heated by exercise or otherwise, it casts off its serum plentifully by sweat, which may be effected by the swift motion of the blood through the glandules of the skin, where its plentiful streams being straitened and constipated into a liquor, force their way through those emunctories, which at other times transmit only insensible vapours. Some such effect may be wrought upon the blood, by the heat of the kidneys. Certain it is, that the humours, excerned by sweat and urine are near akin, if not the same; and therefore it is worthy the consideration, whether there might not be some use made of sweating in a suppression of urine. But I digress too far.

I shall only add to this particular, that because the design of nature in collecting fat in these places, is for the fore-mentioned use; it hath for the effecting thereof fitted the vessels there with pores or passages proper for the separation and transmission of it.

I should now proceed to treat of the generation and formation of the fœtus in the womb; but that is a subject too difficult for me to handle; the body of man and other animals being formed in the dark recesses of the matrix, or as the psalmist phrases it, Psal. cxxxix. 14. 'made in secret, and curiously wrought in the lowest parts of the earth.' This work is so admirable and unaccountable, that neither the atheists nor mechanic philosophers have attempted to declare the manner and process of it; but have (as I noted before) very cautiously and prudently broke off their systems of natural philosophy here, and left this point untouched; and those accounts which some of them have attempted to give of the formation of a few of the parts, are so excessively absurd and ridiculous, that they need no other confutation than ha, ha, he. And I have already farther shewn, that it seems to me impossible, that matter divided into as minute and subtle parts as you will or can imagine, and those moved according to what catholic laws soever can be devised, should, without the presidency and direction of some intelligent agent, by the mere agitation of a gentle heat, run itself into such a curious machine, as the body of man is.

Yet must it be confest, that the seed of animals is admirably qualified to be fashioned and formed by the plastic nature into an organical body, containing the principles or component particles of all the several homogeneous parts thereof; for indeed every part of the body seems to club and contribute to the seed, else why should parents that are born blind or deaf, or that want a finger or any other part, or have one superfluous, sometimes generate children that have the same defects or imperfections; and yet (which is wonderful) nothing of the body or grosser matter of the seed comes near the first principle of the fœtus, or in some so much as enters the womb, but only some contagious vapour or subtle effluviums thereof; which seems to animate the gemma or cicatricula of the egg contained in the female ovary, before it passes through the tubes or cornua into the uterus. How far the animalcules observed in the seed of males, may contribute to generation, I leave to the more sagacious philosophers to inquire; and shall here content myself with referring the reader to the several letters published by M. Lewenhoeck.

But to what shall we attribute the fœtus's likeness to the parents, or omitting them, to the precedent progenitors, as I have observed some parents that have been both black-haired, to have generated most red-haired children, because their ancestors' hair hath been of that colour; or why are twins so often extremely alike? whether is this owing to the efficient, or to the matter?

Those effluvia we spake of in the male seed, as subtile as they are, yet have they a great, if not the greatest, stroke in generation, as is clearly demonstrable in a mule, which doth more resemble the male parent, that is the ass, than the female, or horse. But now why such different species should not only mingle together, but also generate an animal, and yet that that hybridous production should not again generate, and so a new race be carried on; but nature should stop here and proceed no farther, is to me a mystery, and unaccountable.

One thing relating to generation I cannot omit; that is, the construction of a set of temporary parts (like scaffolds in a building), to serve a present end, which are afterward laid aside, afford a strong argument of counsel and design. Now for the use of the young during its inclosure in the womb there are several parts formed, as the membranes inveloping it, called the secundines, the umbilical vessels, one vein and two arteries, the urachus, to convey the urine out of the bladder, and the placenta uterina; part whereof fall away at the birth, as the secundines and placenta, others degenerate into ligaments, as the urachus, and part of the umbilical vein: besides which, because the fœtus during its abode in the womb hath no use of respiration by the lungs, the blood doth not all, I may say not the greatest part of it, flow through them; but there are two passages or channels contrived, one called the foramen ovale, by which part of the blood brought by the vena cava passeth immediately into the left ventricle of the heart, without entering the right at all; the other is a large arterial channel passing from the pulmonary artery immediately into the aorta, or great artery, which likewise derives part of the blood thither, without running at all into the lungs: these two are closed up soon after the child is born, when it breathes no more (as I may so say) by the placenta uterina, but respiration by the lungs is needful for it. It is here to be noted, that though the lungs be formed so soon as the other parts, yet during the abode of the foetus in the womb, they lie by as useless. In like manner I have observed, that in ruminating creatures the three foremost stomachs, not only during the continuance of the young in the womb, but so long as it is fed with milk, are unemployed and useless, the milk passing immediately into the fourth.

Another observation I shall add concerning generation, which is of some moment, because it takes away some concessions of naturalists, that give countenance to the atheists, fictitious and ridiculous account of the first production of mankind, and other animals, viz. that all sorts of insects, yea, and some quadrupeds too, as frogs and mice, are produced spontaneously. My observation and affirmation is, that there is no such thing in nature as equivocal or spontaneous generation, but that all animals, as well small as great, not excluding the vilest and most contemptible insect, are generated by animal parents of the same species with themselves; that noble Italian virtuoso, Francisco Redi, having experimented that no putrified flesh (which one would think were the most likely of any thing) will of itself, if all insects be carefully kept from it, produce any: the same experiment I remember Doctor Wilkins, late bishop of Chester, told me had been made by some of the Royal Society. No instance against this opinion doth so much puzzle me, as worms bred in the intestines of man and other animals. But seeing the round worms do manifestly generate, and probably the other kinds too, it is likely they come originally from seed, which how it was brought into the guts, may afterward possibly be discovered. Moreover, I am inclinable to believe, that all plants too, that themselves produce seed (which are all but some very imperfect ones, which scarce deserve the name of plants), come of seeds themselves. For that great naturalist Malpighius, to make experiment whether earth would of itself put forth plants, took some purposely digged out of a deep place, and put it into a glass vessel, the top whereof he covered with silk many times doubled and strained over it, which would admit the water and air to pass through, but exclude the least seed that might be wafted by the wind; the event was, that no plant at all sprang up in it. Nor need we wonder how in a ditch, bank, or grass-plat newly digged, or in the fen-banks in the Isle of Ely, mustard should abundantly spring up, where in the memory of man none hath been known to grow, for it might come of seed which had lain there more than a man's age : some of the ancients mentioning some seeds that retain their fecundity forty years. And I have found in a paper received from a friend, but whom I have forgotten, that melon-seeds after thirty years, are best for raising of melons. As for the mustard that sprung up in the Isle of Ely, though there had never been any in that country, yet might it have been brought down in the channels by the floods, and so being thrown up the banks, together with the earth, might germinate and grow there.

And indeed a spontaneous generation of animals and plants, upon due examination, will be found to be nothing else than a creation of them. For after the matter was made, and the sea and dry land separated, how is the creation of plants and animals described but by a commanding, that is, effectually causing the waters and earth to produce their several kinds without any seed? Now creation being the work of Omnipotency, and incommunicable to any creature, it must be beyond the power of nature, or natural agents, to produce things after that manner. And as for God Almighty, he is said to have rested from his work of creation after the seventh day. But if there be any spontaneous generation, there was nothing done at the creation, but what is daily done; for the earth and water produced animals then without seed, and so they do still.

Because some, I understand, have been offended at my confident denial of all spontaneous generation, accounting it too bold and groundless, I shall a little enlarge upon it, and give my reasons, in order to their satisfaction.

First, then I say, such a spontaneous generation seems to me to be nothing less than a creation. For, creation being not only a production of a thing out of nothing, but also out of indisposed matter, as may be clearly inferred from the Scripture, and is agreed by all divines; this spontaneous generation, being such a production, wherein doth it differ from creation? or what did God Almighty do at the first creation of animals and plants, more than what (if this be true) we see every day done? To me, I must confess, it seems almost demonstrable, that whatever agent can introduce a form into indisposed matter, or dispose the matter in an instant, must be superior to any natural one, not to say omnipotent.

Secondly, those who have with the greatest diligence and application considered and searched into this matter, as those eminent virtuosi, Marcellus Malpighius, Franciscus Redi, John Swammerdam, Lewenhoeck, and many others, are unanimously of this opinion, save that Franc Redi would except such insects as are bred in galls and some other excrescencies of plants. Now their authority weighs more with me, than the general vogue, or the concurrent suffrages of a thousand others, who never examined the thing so carefully and circumspectly as they have done, but run away with the cry of the common herd of philosophers.

First of all, Dr. Swammerdam, who hath been, to the best purpose of any man I know of, busied in searching out and observing the nature of all insects in general; all in general, I say, for (as to one particular insect, to wit, the silk-worm, I must except Seignior Malpighi; and to one genus of them, to wit, spiders, Dr. Lister;) in his General History of Insects, written in Low Dutch, and translated into French, p. 47. hath these words, 'Nous disons, que il ne se fait dans toute la nature aucune generation par accident,' &c. We affirm, that there is not in all nature any accidental [or spontaneous] generation, but all come by propagation; wherein chance hath not the least part or interest. And in p. 159, speaking of the generation of insects out of plants, in contradiction I suppose to Seignior Redi, he saith, 'Nous croyons absolument,' &c. We do absolutely believe, that it is not possible to prove by experience, that any insects are engendered out of plants : but on the contrary, we are very well informed and assured, that these little animals are not shut up in or enclosed there for any other reason than to draw thence their nourishment. It is true, indeed, that by a certain, constant, and immutable order of nature we see many sorts of insects affixed to particular species of plants and fruits, to which the respective kinds fasten themselves as it were by instinct. But we are to know, that they all come of the seed of animalcules of their own kind, that were before laid there. For these insects do thrust their seed or eggs so deep into the plants, that they come to be afterward as it were united with them, and the aperture or orifice by which they entered quite closed up and obliterated; the eggs being hatched and nourished within. We have often found the eggs of insects so deeply sunk into the tender buds of trees, that without hurting of them it was impossible to draw them out. Many instances he produces in several sorts of insects making their way into plants, which though they be well worth the reading, are too long to transcribe.

Secondly, that great and sagacious naturalist, and most accurate examiner of these things, Seignior Malpighi, in his Treatise of Galls, under which name he comprehends all preternatural and morbose tumours and excrescencies of plants, doth demonstrate in particular, that all such warts, tumours, and excrescencies, where any insects are found, are excited or raised up either by some venenose liquor, which together with their eggs such insects shed upon the leaves, or buds, or fruits of plants, or boring with their terebræ instill into the very pulp of such buds or fruits; or by the contagious vapour of the very eggs themselves producing a mortification or syderation in the parts of plants on which they are laid; or lastly, by the grubs or maggots hatched of the eggs laid there, making their way with their teeth into the buds, leaves, or fruit, or even the wood itself, of such plants on which their eggs were laid. So at last he concludes, 'Erunt itaque gallæ et reliqui plantarum tumores morbosæ excrescentiæ, vi depositi ovi a turbata plantarum compage et vitiato humorum motu excitatæ, quibus inclusa ova et animalcula velut in utero foventur et augentur, donec manifestatis firmatisque propriis partibus, quasi exoriantur novam exoptantia auram.' We conclude therefore, that galls and other tumours of plants are nothing else but morbose excrescencies, raised up by the force of the egg there laid, disturbing the vegetation and temper of the plants, and perverting the motion of their humours and juices; wherein the enclosed eggs and animalcules are cherished, nourished, and augmented, till their proper parts being manifested, explicated, and hardened or strengthened, they are as it were new-born, affecting to come forth into the open air. In the same treatise he describes the hollow instrument (terebra he calls it, and we may English it 'piercer') wherewith many flies are provided, proceeding from the womb, with which they perforate the tegument of leaves, fruits, or buds, and through the hollow of it inject their eggs into the holes or wounds which they have made, where in process of time they are hatched and nourished. This he beheld one of these insects doing, with his own eyes, in the bud of an oak; the manner whereof he describes p. 47. which I shall not transcribe; only take notice, that when he had taken off the insect, he found in the leaf very little and diaphanous eggs, exactly like to those which yet remained in the tubes of the fly's womb.' He adds farther, that it is probable that there may be eggs hidden in divers parts of plants, whereof no footstep doth outwardly appear, but the plant remains as entire, and thrives as well as if there were no insect there: nay, that some may be hidden and cherished in dry places (not wanting any humour to feed them) as in sea-wood, yea, in earthen-vessels, and marbles themselves.

Indeed to me it seems unreasonable that plants, being of a lower form or order of being, should produce animals; for either they must do it out of indisposed matter; and then such production would amount to a creation; or else they must prepare a fit matter, which is to act beyond their strength, there being required to the preparation of the sperm of animals a great apparatus of vessels, and many secretions, concoctions, reflexions, digestions, and circulations of the matter, before it can be rectified and exalted into so noble a liquor: and besides, there must be an egg too, for we know *ex ovo omnia*, to the perfection whereof there are as many vessels, and as long a process required. Now in plants there are no such vessels, and consequently no such preparation of eggs or sperm, which are the necessary principles of animals.

Thirdly, that worthy author of our own country; I mean Dr. Lister, in his Notes upon 'Goedartius Insect.' no. 16. p. 47, hath these words, ' Non enim inducor ut credam, hoc, vel aliud quodvis animal, modo quodam spontaneo e planta produci, et alii causæ cuicunque originem suam debere quam parenti animali:' i. e. ' I cannot be persuaded or induced to believe, that this or any other animal is (or can be) produced out of a plant in a spontaneous manner, or doth owe its original to any other cause whatever than an animal parent of its own kind.' And in his third note upon Insect. no. 49. these, ' Quod spontaneam erucæ hujus aliorumque insectorum generationem pro parte negativa jam sententiam meam tradidi, &c.' As to the spontaneous generation of this eruca and other insects, I have already delivered my opinion for the negative. This is most certain, that these cossi are produced of eggs laid by animal parents: it is also alike clear, that these diminutive caterpillars are able by degrees to pierce or bore their way into a tree; which very small holes, after they are fully entered, do perchance grow together and quite disappear; at least become so small, that they are not to be discerned, unless by Lynceus's eyes. Add moreover, that perchance they undergo no transformation, but continue under the vizard of [erucæ] caterpillars for many years, which doth very well accord with my observations. Moreover, that this caterpillar [eruca] is propagated by animal parents, to wit, butterflies, after the common origination of all caterpillars. In all this I fully consent with the Doctor; only crave leave to differ in his attributing to them the name of cossi, which were accounted by the ancients a delicate morsel, and fed for the table for I take those to have been the hexapods from which the greater sort of beetles come; for that that sort of hexapods are at this day eaten in our American plantations, as I am informed by my good friend Dr. Hans Sloane, who also presented me with a glass of them, preserved in spirit of wine.

Having lately had an opportunity more curiously to view and examine the great flesh-coloured thin-haired English caterpillar (which is so like that sent me by Dr. Sloane that it differs little but in magnitude, which may be owing to the climate), I observed that it had a power of drawing its eight hind-legs or stumps so far up into its body that they did altogether disappear, so that the creature seemed to want them, and of thrusting them out again at pleasure: whereupon I conjectured that the insect of Jamaica sent me by the Doctor (which I took to be the cossus or hexapod previous to some large beetle) had likewise the same power of drawing up its hind-legs, so that though to appearance it wanted them, yet really it did not so, but had only drawn them up, and hid them in its body when it was immersed in the spirit of wine, and consequently was not the hexapod of a beetle, but an eruca, like to, or indeed specifically the same with, that of our own country by me observed; and being eaten at this day by the inhabitants of Jamaica, in all likelihood the same with the cossus of the ancient Romans, which was fed for the table, as Pliny assures us: especially if we consider that Dr. Lister found this eruca in the body of an oak newly cut down, and sawn in pieces, on which tree Pliny saith they feed. Thus much I thought fit to add, to do Dr. Lister and the truth right, by retracting my former conjecture concerning the cossi.

3. My third argument against spontaneous generation, is, because there are no arguments or experiments, which the patrons of it do or can produce, which do clearly evince it. For the general and vulgar opinion, that the heads of children, or the bodies of those that do not change their linen, but wear that which is sweaty and sordid, breed lice; or that cheese of itself breeds mites or maggots, I deny, and look upon it as a great error and mistake; and do affirm,

that all such creatures are bred of eggs laid in such sordid places by some wandering louse, or mite, or maggot. For such places being most proper for the hatching and exclusion of their eggs, and for the maintenance of their young, nature hath endued them with a wonderful acuteness of scent and sagacity, whereby they can, though far distant, find out and make towards them. And even lice and mites themselves, as slow as they seem to be, can to my knowledge, in no long time march a considerable way to find out a convenient harbour for themselves.

Here, by the by, I cannot but look upon the strange instinct of this noisome and troublesome creature the louse, of searching out foul and nasty clothes to harbour and breed in, as an effect of divine providence, designed to deter men and women from sluttishness and sordidness, and to provoke them to cleanliness and neatness. God himself hateth uncleanness, and turns away from it, as appears by Deut. xxiii. 12—14. But if God requires and is pleased with bodily cleanliness, much more is he so with the pureness of the mind : ' Blessed are the pure in heart, for they shall see God ;' Matt. v. 10.

As for the generation of insects out of putrid matter, the experiments of Franciscus Redi, and some of our own virtuosi, give me sufficient reason to reject it. I did but just now mention the quick scent that insects have, and the great sagacity in finding out a proper and convenient harbour or matrix, to cherish and hatch their eggs, and feed their young : they are so acted and directed by nature, as to cast their eggs in such places as are most accommodate for the exclusion of their young, and where there is food ready for them so soon as they be hatched : nay, it is a very hard matter to keep off such insects from shedding their seed in such proper places. 'Indeed, if an insect may be thus equivocally generated, why not sometimes a bird? a quadruped, a man, or even an universe? or why no new species of animal now and then?' as my learned friend Dr. Tancred Robinson very well argues in his Letters: 'for there is as much art shewn in the formation of those as of these.'

A fourth and most effectual argument against spontaneous generation is, that there are no new species produced, which would certainly now and then, nay very often, happen, were there any such thing. For in such pretended generations, the generant or active principle is supposed to be the sun, which being an inanimate body cannot act otherwise than by his heat; which heat can only put the particles of the passive principle into motion. The passive principle is putrid matter; the particles whereof cannot be conceived to differ in any thing but figure, magnitude, and gravity. Now the heat putting these particles in motion, may indeed gather together those which are homogeneous or of the same nature, and separate those that are heterogeneous or of a different; but that it should so situate, place, and connect them, as we see in the bodies of animals, is altogether inconceivable; which if it could, yet that it should always run them into such a machine as is already extant, and not often into some new-fashioned one, such as was never seen before, no reason can be assigned or imagined. This the Epicurean poet Lucretius was so sensible of, that he saw a necessity of granting seeds or principles to determine the species. For, saith he, if all sorts of principles could be connected,

> Vulgo fieri portenta videres,
> Semiferas hominum species exsistere, et altos
> Interdum ramos egigni corpore vivo;
> Multaque connecti terrestria membra marinis;
> Tum flammam retro spirantes ore chimæras
> Pascere naturam per terras omniparenteis:
> Quorum nil fieri manifestum est, omnia quando
> Seminibus certis, certa genetrice, creata,
> Conservare genus crescentia posse videmus, &c.

That is,

> ————Thence would rise
> Vast monsters, nature's great absurdities;
> Some things half beast, half man, and some would grow
> Tall trees above and animals below;
> Some join'd of fish, and beasts, and everywhere
> Frightful chimeras breathing flames appear.
> But since we see no such, and things arise
> From certain seeds of certain shape and size,
> And keep their kind as they increase and grow;
> There's some fixed reason why it should be so.

The raining of frogs and their generation in the clouds, though it be attested by many and great authors, I look upon as utterly false and ridiculous. It seems to me no more likely that frogs should be engendered in the clouds, than Spanish gennets begotten by the wind; for that hath good authors too. And he that can swallow the raining of frogs, hath made a fair step towards believing, that it may rain calves also; for we read that one fell out of the clouds in Avicen's time. Nor do they much help the matter who say, that those frogs that appear sometimes in great multitudes after a shower, are not indeed engendered in the clouds, but coagulated of a certain sort of dust commixed and fermented with rain-water; to which hypothesis Fromondus adheres.

But let us a little consider the generation of frogs in a natural way. 1. There are two different sexes, which must concur to their generation. 2. There is in both a great apparatus of spermatic vessels, wherein the nobler and more spirituous part of the blood is by many digestions, concoctions, reflexions, and circulations exalted into that generous liquor we call sperm; and likewise for the preparing of the eggs. 3. There must be a copulation of the sexes, which I rather mention, because it is the most remarkable in this, that ever I observed in any animal. For they continue in 'complexu venereo,' at least a month indesinently; the male all that while resting on the back of the female, clipping and embracing her with his legs about the neck and body, and holding her so fast, that if you take him out of the water, he will rather bear her whole weight, than let her go. This I observed in a couple kept on purpose in a vessel of water, by my learned and worthy friend Mr. John Nid, fellow of Trinity College, long since deceased. After this the spawn must be cast into water, where the eggs lie in the midst of a copious jelly, which serves them for their first nourishment for a considerable while. And at last the result of all is not a perfect frog, but a tadpole without any feet, and having a long tail to swim withal; in which form it continues a long time, till the limbs be grown out, and the tail fallen away, before it arrives at the perfection of a frog.

Now if frogs can be generated spontaneously in the clouds out of vapour, or upon the earth out of dust and rain-water, what needs all this ado? To what purpose is there such an apparatus of vessels for the elaboration of the sperm and eggs; such a tedious process of generation and nutrition? This is but an idle pomp. The sun (for he is supposed to be the equivocal generant or efficient by these philosophers) could have dispatched the business in a trice: give him but a little vapour, or a little dry dust and rain water, he will produce you a quick frog, nay a whole army of them, perfectly formed, and fit for all the functions of life in three minutes, nay, in the hundredth part of one minute, else must some of those frogs that were generated in the clouds fall down half formed and imperfect; which I never heard they did: and the process of generation have been observed in the production of frogs out of dust and rain-water, which no man ever pretended to mark or discern. But that there can be no frogs generated in the clouds may farther be made appear, first, from the extreme cold of the middle region of the air, where the vapours are turned into clouds, which is not at all propitious to generation. For did not so great men as Aristotle and Erasmus report it, I could hardly be induced to believe, that there could be one species of insects generated in snow. 2. Because if there were any animals engendered in the clouds, they must needs be maimed and dashed in pieces by the fall, at least such as fell in the highways and upon the roofs of houses; whereas we read not of any such broken or imperfect frogs found any where. This last argument was sufficient to drive off the learned Fromondus from the belief of their generation in the clouds; but the matter of fact he takes for granted, I mean the spontaneous generation of frogs out of dust and rain-water, from an observation or experiment of his own at the gates of Tournay in Flanders, to the sight of which spectacle he called his friends who were there present, that they might admire it with him. 'A sudden shower,' saith he, 'falling upon the very dry dust, there suddenly appeared such an army of little frogs, leaping about every where upon the dry land, that there was almost nothing else to be seen. They were also all of one magnitude and colour: neither did it appear out of what lurking places [latibula] so many myriads could creep out and suddenly discover themselves upon the dry and dusty soil, which they hate.' But saving the reverence due to so great a man, I doubt not but they did all creep out of their holes and coverts, invited by the agreeable vapour of the rain-water. This, however unlikely it may seem, is a thousand times more probable, than their instantaneous and undiscernible generation out of a little dry dust and rain-water, which also cannot have any time to mix and ferment together; which is the hypothesis he adheres to. Nay, I affirm, that it is not at all improbable; for he that shall walk out in summer nights when it begins to grow dark, may observe such a multitude of great toads and frogs crawling about in the highways, paths, and avenues to houses, yards, and walks of gardens and orchards, that he will wonder whence they came, or where they lurked all the winter, and all the day-time, for that then it is a rare thing to find one.

To which add, that in such frogs as we are speaking of, Monsieur Perault hath upon dissection often found the stomach full of meat, and the intestines of excrement; whence he justly concludes, that they were not then first formed, but only appeared of a sudden; which is no great wonder, since upon a shower after a drought, earth-worms and land-snails innumerable come out of their lurking-places in like manner.

In confirmation of what I have here written against the spontaneous generation of frogs either in the clouds out of vapour, or on the earth out of dust and rain-water commixed; endeavouring to prove by force of argument that there is no such thing, I have lately received from my learned and ingenious friend Mr. William Derham, rector of Upminster, near Rumford, in Essex, a relation parallel to that of Fromondus concerning the sudden appearance of a vast number of frogs, after a shower or two of rain, marching across a sandy way, that before the rain was very dusty; and giving an account, where in all likelihood they were generated by animal parents of their own kind, and whence they did proceed. The whole narrative I shall give the reader in his own words.

'Some years ago as I was riding forth one afternoon in Berks, I happened upon a prodigious multitude creeping across the way. It was a sandy soil, and the way had been full of dust, by reason of a dry season that then was. But, an hour or two before, a refreshing fragrant shower or two of rain had laid the dust. Whereupon what I had heard or read of the raining of frogs immediately came to my thoughts; as it easily might do, there being probably as good reason then for me, as I believe any ever had before, to conclude that these came from the clouds, or were instantaneously generated. But being prepossessed with the contrary opinion, viz. That there was no equivocal generation, I was very curious in inquiring whence this vast colony might probably come: and upon searching, I found two or three acres of land covered with this black regiment, and that they all marched the same way towards some woods, ditches, and such like cool places in their front, and from large ponds in their rear. I traced them backwards, even to the very side of one of the ponds. These ponds in spawning time always used to abound much with frogs, whose croaking I have heard at a considerable distance; and a great deal of spawn I have found there.

'From these circumstances I concluded, that this vast colony was bred in those ponds from whenceward they steered their course: that after their incubation (if I may so call it) or hatching by the sun, and their having passed their tadpole-state; they had lived (till that time of their migration) in the waters, or rather on the shore, among the flags, rushes, and long grass: but now being invited out by the refreshing showers, then newly fallen, which made the earth cool and moist for their march, that they left their old 'latibula,' where perhaps they had devoured all their proper food; and were now in pursuit of food, or a more convenient habitation.

'This I think not only reasonable to be concluded, but withal so easy to have been discovered by any inquisitive observer, who in former times met with the like appearance, that I cannot but admire that such sagacious philosophers as Aristotle, Pliny, and many others since, should ever imagine frogs to fall from the clouds, or be any way instantaneously or spontaneously generated, especially considering how openly they act their coition, produce spawn, this spawn tadpoles, and tadpoles frogs.

'Neither in frogs only, but also in many other creatures, as lice, flesh-flies, silk-worms, and other papilios, a uniform regular generation was very obvious, which is an argument to me of a strange prepossession of fancy in the ages since Aristotle, not to say of carelessness and sloth.' So far Mr. Derham.

In like manner, doubtless, Fromondus, had he made a diligent search, might have found out the place where those myriads of frogs, observed by him about the gates of Tournay, were generated, and whence they did proceed.

As for the worms and other animals bred in the intestines of man and beast, I have declared myself not to be satisfied of the way and means, how their seeds come to be conveyed into those places; but yet that their generation is analogous to that of other creatures of those kinds I doubt not. The constancy to their species; their exact agreement and perpetual similitude in the shape and figure of their bodies, and all their parts, their consistence, temper, motion, and other accidents, are to me little less than a demonstration, that they are not the effects of chance, but the products of a settled and spermatic principle. I am at present, till better informed, of opinion, that their eggs are swallowed with the meat we eat; and I am the rather induced to think so, because children in their first infancy, and as long as they are constantly confined to a milk diet, are seldom troubled with them.

After this was written, I received a letter from my often remembered ingenious friend Dr. Tancred Robinson, referring to this matter, part whereof I shall transcribe, as being very pertinent, instructive, and consonant to my own thoughts; 'I think it may be proved, that the vast variety of worms found in almost all the parts of different animals, as well terrestrial as aquatic, are taken into their respective bodies by meats and drinks, and there either lie still for some time, or else grow and alter by change of place and food, (not specifically but accidentally in magnitude, colour, figure of some parts, or the like.) We know as yet but little of the numerous insects bred in water, or indeed of those in roots, leaves, buds, flowers, fruits, and seeds, which we are continually swallowing; and these too all vary according to climate. (That is, the same species of roots, leaves, &c. do in different climates produce many different species of insects, though some there be common to all.) The long slender worms, as small as hairs, that breed between the skin and flesh in the isle of Ormuz and in India; which are generally twisted out upon sticks or rollers, and often break in the operation, are without doubt taken in by the water they drink in those regions, as I could prove by many and good experiments, had I time. They who have leisure, may find them in the collections of voyages and travels, especially in Monsieur Thevenot. By this explication we may give a better account of the vomitings up of tadpoles, snails, and other animals, recorded in medical histories, than by any hypothesis of equivocal generation: as to insects found in stinking flesh, or rotten vegetables, I could never observe or find any of them different from those parent insects, which hover about, or feed upon such bodies.'

If any shall object, the infinite multitudes of animalcules discovered in pepper-water, and desire an account of their generation; to him I shall say, that it is probable, that some few of these animals may be floating in all waters, and that finding the particles of pepper swimming in the water very proper for the cherishing and excluding of their eggs by reason of their heat, or some other unknown and specific quality, they may fasten their eggs to them, and so there may be a sudden breed of infinite swarms of them. But these being not to be discerned by the most piercing and lyncean sight without the assistance of a microscope, I leave the manner of their generation to future discovery.

No less difficult it is to give an account of the original of such insects as are found, and seem to be bred in the bodies of others of different kinds. Out of the sides and back of the most common caterpillar, which feeds upon cabbage, cole-wort, and turnip-leaves, which we have described in the catalogue of Cambridge-plants, we have seen creep out small maggots to the number sometimes of threescore or more, which so soon as ever they came forth, began to weave themselves silken cases of a yellow shining colour, wherein they changed, and after some time came out thence in the form of small flies with four wings; for a full description and history whereof, I shall refer the reader to the forementioned catalogue. The like I have also observed in other caterpillars of a different kind, which have produced no lesser number of maggots, that in like manner immediately made themselves up in cases. Others instead of changing into aurelias, as in the usual process of nature they ought to do, have turned into one, two, three, or more flesh-fly cases, at least contained such cases within them, out of which, after a while, were excluded flesh-flies. Other caterpillars, as that called the solitary maggot, found in the dry heads of teasel, by a dubious metamorphosis sometimes changed into the aurelia of a butterfly, sometimes into a fly-case. You will say, How comes this to pass? Must we not here necessarily have recourse to a spontaneous generation? I answer, no: the most that can be inferred from hence is, a transmutation of species; one insect may instead of generating another of its own kind, beget one or more of a different. But I can by no means grant this. I do believe that these flies do either cast their eggs upon the very bodies of the forementioned caterpillars, or upon the leaves on which they feed, all in a string: which there hatching, eat their way into the body, where they are nourished till they be come to their full growth. Or it may be, the fly may, with the hollow and sharp tube of her womb, punch and perforate the very skin of the eruca, and cast her eggs into its body. So the ichneumon will convey her eggs into caterpillars.

The discovery of the manner of the generation of these sorts of insects I earnestly recommend to all ingenious naturalists, as a matter of great moment. For if this point be but cleared, and it be demonstrated that all creatures are generated univocally by parents of their own kind, and that there is no such thing as spontaneous generation in the world, one main prop and support of atheism is taken away, and their strongest hold demolished: they cannot then exemplify their foolish hypothesis of the generation of man and other animals at first by the like of frogs and insects at this present day.

It will be farther objected, that there have live toads been found in the midst of timber-trees: nay, of stones when they have been sawn under.

To this I answer, that I am not fully satisfied of the matter of fact. I am so well acquainted with the credulity of the vulgar, and the delight they and many of the better sort too, have in telling of wonders and strange things, that I must have a thing well attested, before I can give a firm assent to it.

Since the writing hereof the truth of these relations of live toads found in the midst of stones, hath been confirmed to me by sufficient and credible eye-witnesses, who have seen them taken out. So that there is no doubt of the matter of fact.

But yet, suppose it be true, it may be accounted for. Those animals when young and little, finding in the stone some small hole reaching to the middle of it, might, as their nature is, creep into it, as a fit 'latibulum' for the winter, and grow there too big to return back by the passage by which they entered, and so continue imprisoned therein for many years; a little air, by reason of the coldness of the creature, and its lying torpid there, sufficing it for respiration, and the humour of the stone, by reason it lay immovable and spent not, for nourishment. And I do believe, that if those who found such toads, had diligently searched, they might have discovered and traced the way whereby they entered in, or some footsteps of it. Or else there might fall down into the lapideous matter, before it was concrete into a stone, some small toad (or some toad-spawn) which being not able to extricate itself and get out again, might remain there imprisoned till the matter about it were condensed and compacted into a stone. But however it came there, I dare confidently affirm it was not there spontaneously generated. For else either there was such a cavity in the stone before the toad was generated; which is altogether improbable, and 'gratis dictum,' asserted without any ground, or the toad was generated in the solid stone, which is more unlikely than the other, in that the soft body of so small a creature should extend itself in such a prison, and overcome the strength and resistance of such a great and ponderous mass of solid stone.

And whereas the assertors of equivocal generation were wont to pretend the imperfection of these animals, as a ground to facilitate the belief of their spontaneous generation; I do affirm, that they are as perfect in their kind, and as much art shewn in the formation of them, as of the greatest; nay, more too, in the judgment of that great wit and natural historian Pliny:* 'In magnis siquidem corporibus,' saith he, ' aut certe majoribus, facilis officina sequaci materia fuit;

* Lib. xi. cap. 2.

in his tam parvis, atque tam nullis, quæ ratio, quanta vis, quam inextricabilis perfectio?' In the greater bodies the forge was easy, the matter being ductile and sequacious, obedient to the hand and stroke of the artificer, apt to be drawn, formed, or moulded into such shapes and machines, even by clumsy fingers: but in the formation of these, such diminutive things, such nothings, what cunning and curiosity! What force and strength was requisite, there being in them such inextricable perfection!

To what proofs or examples of spontaneous generation may be brought from insects bred in the fruits or excrescences of plants, I have already made answer in my second particular, which contains the testimonies of our best modern naturalists concerning these things.

In my denial of the spontaneous generation of plants, I am not so confident and peremptory; but yet there are the same objections and arguments against it, as against that of animals, viz. because it would be a production out of indisposed matter, and consequently a creation; or if it be said, there is disposed matter, prepared by the earth, or sun, the heat, or whatever other agent you can assign; I reply, this is to make a thing act beyond its strength, that is, an inferior nature, which hath nothing of life in it, to prepare matter for a superior, which hath some degree of life; and for the preparation of which it hath no convenient vessels or instruments. If it could do so, what need of all that apparatus of vessels, preparation of seed, and as I also suppose, distinction of masculine and feminine that we see in plants? I demand farther, whether any of the patrons of spontaneous generation in plants, did ever see any herbs or trees, except those of the grass-leaved tribe, come up without two seed-leaves; which if they never did nor could, it is to me a great argument, that they came all of seed; there being no reason else, why they should at first produce two seed-leaves different from the subsequent. And if all these species (which are far the greatest number) come from seed, there is not the least reason to think that any of the rest come up spontaneously. And this, with what I have written before, may suffice concerning this point.

Whereas I have often written in many places, that such and such plants are spontaneous, or come up spontaneously; I mean no more by that expression, but that they were not planted or sown there industriously by man.

Having spoken of the body of man, and the uses of its several parts and members, I shall add some other observations, giving an account of the particular structure, actions, and uses of some parts either common to whole kinds of animals, or proper to some particular species different from those of man, and of the reason of some instincts and actions of brutes.

First of all, The manner of respiration and the organs serving thereto in various animals, are accommodated to their temper of body and their place and manner of living; of which I have observed in more perfect animals three differences.

1. The hotter animals, which require abundance of spirits for their various motions and exercises, are provided with lungs, which indesinently draw in and expel the air alternately without intermission, and have a heart furnished with two ventricles, because to maintain the blood in that degree of heat, which is requisite to the performance of the actions of all the muscles, there is abundance of air necessary. I shall

not now take notice of the difference that is between the lungs of quadrupeds and birds, how the one is fixed and immoveable, the other loose and moveable; the one perforated, transmitting the air into large bladders, the other enclosed with a membrane.

It is here worth the notice taking, that many animals of this kind, both birds and quadrupeds, will endure and bear up against the extremest rigour of cold that our country is exposed to. Horse, kine, and sheep, as I have experienced, will lie abroad in the open air upon the cold ground during our long winter nights in the sharpest and severest frosts that ever happened with us, without any harm or prejudice at all; whereas one would think, that at least the extremities of their members should be bitten, benumbed, and mortified thereby. Considering with myself by what means they were enabled to do this and to abide and resist the cold, it occurred to my thoughts that the extremities of their toes were fenced with hoofs, which in good measure secured them: but the main thing was, that the cold is, as it were, its own antidote; for the air being fully charged and sated with nitrous or some other sort of particles (which are the great efficients of cold, and no less also the pabulum of fire) when inspired, doth by means of them cause a great accension and heat in the blood (as we see fuel burns rashly in such weather), and so enable it to resist the impressions of the cold for so short a time as its more nimble circulation exposes it thereto, before it comes to another heating. From hence may an account be given why the inhabitants of hot countries may endure longer fasting and hunger than those of colder; and those seemingly prodigious and to us scarce credible stories of the fastings and abstinence of the Egyptian monks be rendered probable.

2. Other animals, which are of a colder temper, and made to endure a long inedia or fasting, and to lie in their holes almost torpid all winter, as all kinds of serpents and lizards, have indeed lungs, but do not incessantly breathe, or when they have drawn in the air necessarily expire it again, but can retain it at their pleasure, and live without respiration whole days together, as was long since experimented by Sir Thomas Brown, M.D. in a frog tied by the foot under water for that purpose by him: this order of creatures have but one ventricle in their hearts; and the whole blood doth not so often circulate through the lungs as it doth through the rest of the body. This manner of breathing is sufficient to maintain in them that degree of heat which is suitable to their nature and manner of living. For to our touch they are always cold even in summer time, and therefore some will then put snakes into their bosom to cool them.

3. Fishes which were to live and converse always in a cold element, the water; and therefore were to have a temper not excelling in heat, because otherwise the constant immediate contact of the water (unless some extraordinary provision were made) could not have been supported by them; that they might not be necessitated continually to be coming up to the top of the water to draw in the air; and for many other reasons that might be alleged, perform their respiration under water by the gills, by which they can receive no more air than is dispersed in the pores of the water which is sufficient to preserve their bodies in that temper of heat that is suitable to their nature and the place wherein

they live. These also have but one ventricle in their hearts.

But now, though this be thus, the great and most wise God, as it were purposely to demonstrate that he is not by any condition or quality of place necessarily determined to one manner of respiration, or one temper of body in fishes; he hath endued the bodies of some of that tribe of aquatic creatures with lungs like viviparous quadrupeds, and two ventricles of the heart, and an ability of breathing like them by drawing in and letting out the open air; so contriving their bodies, as to maintain in the midst of the cold water a degree of heat answerable to that of the forementioned quadrupeds.

Another remarkable thing relating to respiration is the keeping the hole or passage between the arteria venosa and vena cava, called 'foramen ovale, open in some amphibious quadrupeds, viz. the phoca, or vitulus marinus, called in the English sea-calf and seal, and as is generally held, the beaver too. We have already given the reason of the twofold communication of the great blood-vessels in the fœtus or young, so long as it continues in the womb: the one between the two veins entering the heart, by a hole or window; the other between the two arteries, by an arterial channel, extended from the pulmonary artery to the aorta or great artery; which was in brief, to divert the blood from the lungs. The same reason for keeping open this foramen ovale there is in these amphibious creatures; for, 1. The lungs probably being not extended, but emptied of air when they abide long under water, and flaccid, it is not easy for the whole blood every circulation to make its way through them. 2. To maintain that degree of heat and motion in the blood, as is sufficient for them while they are under water, there is not so much air required, as is when they are above. The blood then moving but gently, as doth that of the fœtus in the womb.

Farther, in reference to respiration, it is observed by the Parisian academists, that some amphibious quadrupeds, particularly the sea-calf or seal, hath his epiglottis extraordinarily large in proportion to other animals, it extending half an inch in length beyond the glottis to cover it. I believe the beaver hath the like epiglottis exactly closing the larynx or glottis, and hindering all influx of water; because in one dissected by Wepferus that suffocated itself in the water, there was not a drop of water found in the lungs. It is probable (say they) that this is done more exactly to close the entrance of the aspera arteria, or windpipe, when the animal eats his prey at the bottom of the sea, and to hinder the water from running into his lungs. An elephant (as is observed by Dr. Moulins, I think, in the anatomy of that creature) hath no epiglottis at all, there being no danger of any thing falling into the lungs from eating or drinking, seeing there is no communication between the œsophagus and it. For he thus describes the œsophagus or gullet. The tongue of this creature, saith he, had this peculiar in it, that the passage to the ventricle was through it; for there was a hole near the root of it, and exactly in the middle of that part. Which hole was the beginning of the œsophagus. There was no communication between this and the passage into the lungs, contrary to what we may observe in men, in all quadrupeds and fowl, that ever I had opportunity to dissect. For the 'membrana pituitaria anterior' reached to the very root of the tongue below the œsophagus, and so quite stopped the passage of the air into the mouth. But though there be no danger of meat or drink falling into the lungs; yet were they not sufficiently secured from small animals creeping in there: for though to supply in some measure the want of an epiglottis by lessening the glottis, there grew to the outside of the cartilages called arytenoides, another capable of motion up and down by the help of some muscles that were implanted in it, strong on both sides of the 'aspera arteria,' but on the under side, opposite to that of the œsophagus, very limber, wanting about two inches and a half of coming round the aforesaid cartilages on the upper side, or the next to the œsophagus: yet did not this cartilage so shut up the way against them, but that even a mouse creeping up his proboscis might get into his lungs, and so stifle him. Whence we may guess at the reason why the elephant is afraid of a mouse: and therefore to avoid this danger, this creature (the elephant which this author described) was observed always, when he slept, to keep his trunk (proboscis) so close to the ground, that nothing but air could get in between them. This is a strange sagacity and providence in this animal, or else an admirable instinct.

Again, The Parisian academists observe of the sea-tortoise, that the cleft of the glottis was strait and close. Which exact enclosure, I do rather believe, is to prevent the water from entering into the windpipe, when the tortoises are under water, than to assist the effect of the compression of the air in the lungs, as they would have it. For they make the main reason of respiration, and use of the lungs in this creature to be, to take in and retain air, by the compression and dilatation whereof, made by the muscles, it can raise or sink itself in the water as need requires;

though I do not exclude this. But if this be the main use of the lungs and respiration in this animal, what is in land-animals which have alike conformation of lungs, and manner of respiration; as the cameleon, serpents, and lizards?

But before I dismiss the tortoise, I shall add two notable observations concerning him, borrowed of the said French academists, which seem to argue something of reason in him, and more than a bare instinct. The first is in the land-tortoise, and it is his manner of turning himself, and getting upon his feet again when he is cast upon his back, which they describe in these words; ' At the great aperture of the shell before, there was at the top a raised border, to grant more liberty to the neck and head, for lifting themselves upwards. And this inflection of the neck, is of great use to the tortoises. For it serves them to turn again, when they are upon their backs. And their industry upon this account is very admirable. We have observed in a living tortoise, that being turned upon its back, and not being able to make use of its paws for the returning of itself, because they could only bend towards the belly, it could help itself only by its neck and head, which it turned sometimes on one side, sometimes on the other, by pushing against the ground, to rock itself as in a cradle, to find out the side towards which the inequality of the ground might more easily permit it to roll its shell. For when it had found it, it made all its endeavours on that side.'

The second is in the sea-tortoise, as follows: Aristotle and Pliny have remarked, that when tortoises have been a long time upon the water during a calm, it happens that their shell being dried in the sun, they are easily taken by the fishermen; by reason they cannot plunge into water nimbly enough, being become too light. This shews what equality there ought to be in their equilibrium, seeing so little a change as this, which may happen by the sole drying of the shell, is capable of making it useless. This easiness to be taken at such a time, these academists do not refer merely to the lightness of this creature's body (for he could easily let air enough out of his lungs, to render it heavier than the water, and so enable himself to sink), but to a wonderful sagacity and caution of this animal. For (say they) it is probable that the tortoise, which is always careful to keep himself in this equilibrium, so as other animals are to keep themselves on their legs, in this case by the same instinct, dares not let the air out of his lungs, to acquire a weight which might make him speedily to sink; because he fears that his shell being wet, it should become so heavy, that he being sunk to the bottom of the water, might never have power afterward to reascend. If this may be the reason why he exposes himself to the danger of being taken at such a time, rather than he will descend suddenly to the bottom, it is clear, that he is endued with an admirable providence and foresight, and a power of argumentation.

That nature doth really design the preservation and security of the more infirm creatures, by the defensive armour that it hath given to some of them, together with skill to use it, is, I think, demonstrable in the common hedge-hog, or urchin, and one species of tatou, or armadillo. The hedge-hog hath his back, sides and flanks 'thick set with strong and sharp prickles, and besides, by the help of a muscle, given him for that purpose, is enabled to contract himself into a globular figure, and so to withdraw, enclose, and hide his whole under part, head, belly, and legs (which for the necessities and conveniences of life, must be left destitute of this armour) within his covert or thicket of prickles; so that dogs or other rapacious creatures cannot lay hold upon him, or bite him, without wounding their own noses and mouths. The muscle whereby he is enabled to draw himself thus together, and gather up his whole body like a ball, the Parisian academists describe to be a distinct carnose muscle, extended from the ' ossa innominata' to the ear and nose, running along the back-bone, without being fastened thereto. Olaus Borrichius, in the Danick Transactions, makes it to be an almost circular muscle embracing the 'panniculus carnosus,' of a wonderful fabric, variously extending its ' laciniæ,' or processes, to the feet, tail, and head of the creature.

The other creature, which doth thus contract and draw up itself into a globular or oval figure for its defence, is the second sort of tatou or armadillo, largely described by Marcgrave, lib. 6. cap. 9. by the name of ' Tatu apara,' which is covered on its back and sides, with a strong scaly crust or shell, of a hard or bony substance, jointed like armour or the scales of the tail of a lobster, by four transverse commissures in the middle of the body, connected by tough membranes. When it sleeps (as it doth for the most part in the day-time, going forth to feed in the night), or when one goes about to lay hold on it, gathering up its fore and hind legs as it were to one point, and drawing its ears with its head inward, and bringing its tail to its head, by reason of the forementioned commissures, it bends it back so far, till its head comes to touch its hindpart, and so with its armour gathers itself into a round ball, the lateral extremities of the shell touching one another, and enclosing the body on the sides, and the fore and hind parts coming so near together, that there is nothing to be seen, but the armature of the head and tail, which, like doors, shut up the hole, which the shells of the body left open. This it performs by the action of a notable muscle on each side, of a great length, having the form of the letter x, made up of many fibres, decussating one another longways, by the help whereof it can contract its shell, and hold it contracted with such a mighty force, that he must be a strong man indeed that is able to open it.

Had such a muscle as this, and such an ability of contraction been given to any other creature that was covered with soft hair or fur, there might have been some pretence to fancy, that this was accidental and not designed: but seeing there is not one instance of this kind in nature, it must be great stupidity to believe it, and impudence to assert it. Neither will the atheists' usual κρησφύγετον, or refuge, 'That there were indeed at first such creatures produced, but being obnoxious to those that were strong and rapacious, they were by degrees destroyed, and the race lost,' here help them out: because such a muscle, and faculty of using it to that purpose, might as likely have fallen to the lot or chance of a strong and generous creature, which others dared not approach to hurt, who might for his own disport have thus contracted himself into a ball, of which kind we find none.

I have before mentioned the use assigned by the honourable Mr. Boyle, of famous memory, lately deceased, to the periophthalmium, or nictating membrane in brutes, wherein I could not fully acquiesce as to some quadrupeds, which were in no danger of having their eyes harmed by bushes and prickles, or twigs of trees, since the writing whereof, I have met with a different account of the use of that membrane in the 'Anatomical Descriptions of several Creatures dissected by the Royal Academy of Sciences at Paris,' Englished by Mr. Alexander Pitfeild, p. 249. in the description of the Cassowar. Our opinion (say those academists) is, that the membrane serves to clean the cornea, and to hinder, that by drying, it grow not less transparent. Man and the ape, which are the sole animals wherein we have not found this eye-lid, have not wanted this provision for the cleansing of their eyes, because that they have hands, with which they may, by rubbing their eye-lids, express the humidity which they contain, and which they let out through the ductus lachrymalis: which is known by experience, when the sight is darkened, or when the eyes suffer any pain or itching: for these accidents do cease, when the eyes are rubbed.

But the dissection has distinctly discovered to us the organs which do particularly serve for this use, and which are otherwise in birds, than in man, where the ductus passes not beyond the glandula lachrymalis. For in birds it goes beyond; and penetrating above half-way on the internal eye-lid, it is opened underneath upon the eye: which is evidently done to spread a liquor over the whole cornea, when this eye-lid passes and repasses; as we observed it to do every moment.

The artifice and contrivance of nature for the extending and withdrawing of this curtain of the periophthalmium in birds is admirable; but it is difficult so to express it in words, as to render it intelligible to the reader; for a multitude of words doth rather obscure than illustrate, they being a burden to the memory, and the first apt to be forgotten, before we come to the last. So that he that uses many words for the explaining any subject, doth, like the cuttle-fish, hide himself for the most part in his own ink. And in the description of the figure and manner of the extension and contraction of this membrane, the Parisian academists are constrained to use so many words, that I am afraid few readers' patience and attention will last so long, as to comprehend and carry it away: yet because it is so evident and irrefragable a proof of wisdom and design, I could not omit it. Their words are these: The particularities of the admirable structure of this eye-lid, are such things as do distinctly discover the wisdom of nature, among a thousand others, of which we perceive not the contrivance, because we understand them only by the effects, of which we know not the causes; but we here treat of a machine, all the parts whereof are visible, and which need only to be looked upon, to discover the reason of its motion and action.

This internal eye-lid in birds is a membranous part, which is extended over the cornea, when it is drawn upon it like a curtain by a little cord or tendon; and which is drawn back again into the great corner of the eye to uncover the cornea, by the means of the very strong ligaments that it has, and which in drawing it back towards its origin, do fold it up. It made a triangle when extended, and it had the figure of a crescent when folded up. Its basis (which is its origin) was toward the great corner of the eye, at the edge of the great circle, which the sclerotica forms when it is flatted before making an angle with its anterior part, that is the cornea, which is raised like a hill upon it. The basis, which is the part immoveable, and fastened to the edge of the sclerotica, did take up more than a third part of the circumference of the great circle of the sclerotica, the side of the triangle, which is toward the little corner of the eye, and is moveable, was reinforced with a border, which supplies the place of the tarsus, and which is black in most quadrupeds. This side of the eye-lid, is that which is drawn back into the corner of the eye, by the action of the fibres of the whole eyelid, which parting from its origin, proceed to join themselves to its tarsus.

To extend this eye-lid over the cornea, there were two muscles that were seen, when six were taken away, which served to the motion of the whole eye. We found that the greatest of these two muscles has its origin at the very edge of the great circle of the sclerotica, towards the great corner, from whence the eye-lid takes its original. It is very fleshy in its beginning, which is a large basis, from whence coming insensibly to contract itself by passing under the globe of the eye, like as the eye-lid passes over it, it approaches the optic nerve, where it produces a tendon round and slender, so that it passes through the tendon of the other muscle, which serves for a pully, and which hinders it from pressing the optic nerve upon which it is bent, and makes an angle, to pass through it to the upper part of the eye: and coming out from underneath the eye to insert itself at the corner of the membrane, which makes the internal eye-lid. This second muscle hath its original at the same circle of the sclerotica, but opposite to the first, towards the little corner of the eye, and passing under the eye like the other, goes to meet it, and embraces its tendon, as has been declared.

The action of these two muscles is, in respect to the first, to draw by the means of its cord or tendon the corner of the internal eye-lid, and to extend it over the cornea. As to the second muscle, its action is by making its tendon to approach toward its origin, to hinder the cord of the first muscle, which it embraces, from hurting the optic nerve; but its principal use is, to assist the action of the first muscle. And it is herein that the mechanism is marvellous in this structure, which makes that these two muscles, joined together, do draw much farther than if it had but one. For the inflexion of the cord of the first muscle, which causes it to make an angle on the optic nerve, is made only for this end: and a single muscle with a strait tendon, had been sufficient, if it had power to draw far enough. But the traction which must make the eye-lid extend over the whole cornea, being necessarily great, it could not be done but by a very long muscle; and such a muscle not being able to be lodged in the eye all its length, there was no better way to supply the action of a long muscle, than by that of two indifferent ones, and by bending one of them, to give it the greater length in a little space. Thus far the academists, who themselves reflecting on the length and obscurity of this description, tell us, that the inspection of the figure will serve greatly to the understanding of it, which the novelty of the thing renders obscure in itself; and so I fear it will be to most readers; howbeit in such a work as this, I ought not by any means, as I said before, to leave out such a notable instance, wherein contrivance and design do so clearly and undeniably appear.

The same academists, as I remember, tell us, that they have found by experience, that the aqueous humour of the eye will not freeze; which is very admirable, seeing it hath the perspicuity and fluidity of common water, and hath not been taken notice of, so far as I have heard, to have any eminent quality discoverable either by taste or smell; so that it must be of some singular and ethereal nature: and deserves to be examined and analyzed by the curious naturalists of our times.

The providence of nature is wonderful in a camel, or dromedary, both in the structure of his body, and in the provision that is made for the sustenance of it. Concerning the first, I shall instance only in the make of his foot; the sole whereof, as the Parisian academists do observe, is flat and broad, being very fleshy, and covered only with a thick, soft, and somewhat callous skin, but very fit and proper to travel in sandy-places, such as are the deserts of Afric and Asia. We thought (say they) that this skin was like a living sole, which wore not with the swiftness and the continuance of the march; for which this animal is almost indefatigable. And it may be this softness of the foot, which yields and fits itself to the ruggedness and unevenness of the roads, does render the feet less capable of being worn than if they were more solid.

As to the second, the provision that is made for their sustenance in their continued travels over sandy deserts, the same academists observe, that at the top of the second ventricle (for they are ruminant creatures, and have four stomachs) there were several square holes, which were the orifices of about twenty cavities, made like sacks placed between the two membranes, which do compose the substance of this ventricle. The view of these sacks made us to think, that they might well be the reservatories, where Pliny says, that camels do a long time keep the water, which they drink in great abundance when they meet with it, to supply the wants which they may have thereof in the dry deserts, wherein they are used to travel ; and where, it is said, that those that do guide them are sometimes forced, by extremity of thirst, to open their bellies, in which they do find water.

That such an animal as this, so patient of long thirst, should be bred in such droughty and parched countries, where it is of such eminent use for travelling over those dry and sandy deserts, where no water is to be had sometimes in two or three days' journey, no candid and considerate person but must needs acknowledge to be an effect of providence and design.

Such animals as feed naturally upon flesh, both quadrupeds and birds, because such kind of food is high and rank, do qualify it, the one by swallowing the hair or fur of the beasts they prey upon, the other by devouring some part of the feathers of the birds they gorge themselves with, not electively, but because they cannot, or will not take the pains fully to plume them. And therefore the Parisian academists do rationally refer the death of one of the lions whom they dissected, to the feeding of him with too succulent and delicate meat. For (say they) we know, that sometime before his death he was several months without going out of his den, and that it was hard to make him eat. That for this reason some remedies were prescribed to him, and among others the eating only the flesh of young animals, and those alive. And that those which looked to the beasts of the park of Vicennes, to make this food more delicate, did use a method very extraordinary ; which was, they flead lambs alive, and thus made him eat several; which at the first revived him, creating in him an appetite, and making him brisk. But it is probable that this food engendered too much blood, and which was too subtle for an animal to whom nature had not given the industry of flaying those which he eat. It being credible that the hair, wool, feathers, and scales, which all animals of prey do swallow, are a seasonable and necessary corrective, to prevent their greediness from filling themselves with too succulent a food.

Though I have declared in the beginning of this work, that the means whereby cartilaginous fishes raise and sink themselves in the water, and rest and abide in what depth they please, is not yet certainly known; yet I shall propound a conjecture concerning it, which was first suggested to me by Mr. Peter Dent, late physician in Cambridge, viz. that it is by the help of water which they take in and let out by two holes in the lower part of their abdomen or belly, near the vent, or not far off it. The flesh of this sort of fish being lax and spongy, and nothing so firm, solid, and weighty as that of the bony fishes, and there being a good quantity of air contained in the cavity of their abdomen, they cannot sink in the water without letting in some of it by these holes (the orifices whereof are opened and shut at pleasure by the help of muscles provided for that purpose) into the hollow of their bellies, whereby they preponderate the water and descend; and when they would ascend, by a compression wrought by the muscles of the abdomen, they force out the water again, or at least so much of it as may suffice to give that degree of levity they need or desire. If it be found by experience that the bodies of these fishes without this ballast would naturally float in the water, and that they do really admit water into their bellies, then this conjecture may have some probability or truth in it; otherwise not.

Upon the contemplation and consideration of those various ways and contrivances which nature (I mean the divine wisdom) hath made use of for preparing the chyle, separating the nutritious juice from the grosser parts of the aliment, and the several humours and spirits from the blood, I cannot but admire her great wisdom, art, and curiosity. For she hath not only employed all those methods and devices, which chemists have either learned by imitation of her, or invented of themselves, for analyzing of bodies, separating their parts, the pure from the impure, and extracting their spirits, &c. as maceration in the first stomach or paunch of ruminating creatures, and in the craws of birds; comminution by grinding in the mouths of viparous quadrupeds, and in the gizzards of poultry; fermentation in the stomachs of most terrestrial and all aquatic animals; expression and squeezing, in the omasus of ruminant quadrupeds, and in the intestines of all creatures, by the motion of the midriff and other muscles of the belly forcing the chyle out of the fæces or excrements into the lacteal veins; straining, or percolation, by all the viscera of the body; which are but as so many colanders to separate several juices from the blood; and lastly digestion and circulation in the spermatic parts and vessels, and perhaps also in the brain. I say, it hath not only made use of these operations, but it hath quite outdone the chemists, effecting that by a gentle heat, which they cannot perform without great stress of fire. As for instance, in the stomach of a dog preparing a liquor that dissolves bones; and in the bodies of some insects a liquor which seems to be as highly acid and corrosive as oil of vitriol or spirit of nitre, viz. that which is instilled into the blood when they sting. It is an experiment I have met with in some books, and made myself, that if you put blue-bottles or other blue flowers into a ant-hill, they will presently be stained with red; the reason (which these authors render not) is because the ants thrust in their stings into the flowers, and instil into or drop upon them a small mite of their stinging liquor, which hath the same effect that oil of vitriol would have in changing their colour, which is a sign that both liquors are of the same nature.

Caspar Bartholine hath observed, that where the gullet perforates the midriff, the carneous fibres of that muscular part are inflected and arcuate, as it were a sphincter embracing and closing it fast, by a great providence of nature, lest in the perpetual motion of the diaphragm, the upper orifice of the stomach should gape, and cast out the victuals as fast as it received it. And Peyerus thinks he hath observed, that in ruminating creatures the connexion of the gullet with the diaphragm is far straiter and stronger than in man and other animals, to the end that there should not be more than one morsel forced out at once. For that external sphincter inhibits a too great dilatation of the gullet, and doth, as it were, measure out the morsels, and fit them to the capacity of the œsophagus.

I shall conclude with a notable relation of Galen's, lib. 6. ' de locis affectis,' cap. 6. concerning a kid taken by him alive out of the dam's belly, and nourished and brought up.

Ἡ διαπλάσασά τε καὶ τελειώσασα φύσις εἰργάσατο χωρὶς διδασκαλίας ἐπὶ τὴν οἰκείαν ἐνέργειαν ἔρχεσθαι· καὶ βάσανόν γε τούτου μεγίστην ἐποιησάμην ποτὲ θρέψας ἔριφον, ἄνευ τοῦ θεάσασθαί ποτε τὴν κυήσασαν· αἶγας γὰρ ἐγκύμονας ἀνατεμὼν ἕνεκα τῶν ἐζητημένων θεωρημάτων τοῖς ἀνατομικοῖς ἀνδράσι περὶ τῆς κατὰ τὸ κυούμενον οἰκονομίας, εὑρών ποτε γενναῖον τὸ ἔμβρυον, ἀπέλυσα μὲν τῆς μήτρας ὥσπερ εἰώθαμεν· ἁρπάσας δὲ πρὶν θεάσασθαι τὴν κυήσασαν εἰς οἶκον μέν τινα κομίσας κατέθηκα, πολλὰ μὲν ἔχοντα λεκάνια· τὸ μὲν οἴνου, τὸ δὲ ἐλαίου, τὸ δὲ μέλιτος, τὸ δὲ γάλακτος, ἢ ἄλλου τινὸς ὑγροῦ πλῆρες, οὐκ ὀλίγα δι' ἄλλα τῶν Δημητρείων καρπῶν, ὥσπερ δὲ καὶ τῶν ἀκροδρύων· ἐθεασάμεθα δὲ τὸ ἔμβρυον ἐκεῖνο, πρῶτον μὲν βαδίζον τοῖς ποσὶν, ὥσπερ ἀκηκοὼς ἕνεκα βαδίσεως ἔχειν τὰ σκέλη· δεύτερον δὲ ἀποσειόμενον τὴν ἐκ τῆς μήτρας ὑγρότητα, καὶ τρίτον ἐπὶ τούτῳ κυσάμενον ἑνὶ τῶν ποδῶν τὴν πλευρὰν, εἶτ' ὀσμώμενον εἴδομεν αὐτὸ τῶν κειμένων κατὰ τὸν οἶκον ἑκάστου, ὡς δὲ πάντων ὠσμᾶτο, τοῦ γάλακτος ἀπερρόφησεν, ἐν ᾧ καὶ ἀνεκεκράγαμεν ἅπαντες, ἐναργῶς ὁρῶντες ὅπερ Ἱπποκράτης ἔφη, φύσεις ζῴων ἀδίδακτοι. Καὶ τοίνυν καὶ ἀνεθρέψαμεν ἐκεῖνο τὸ ἐρίφιον, εἴδομέν τε προσφερόμενον ὕστερον, οὐ τὸ γάλα μόνον, ἀλλὰ καὶ ἄλλα τινὰ τῶν κειμένων· ὄντος δὲ τοῦ καιροῦ καθ' ὃν ἐξῃρέθη τῆς μητρὸς ὁ ἔριφος, ἐγγὺς τῆς ἐαρινῆς ἰσημερίας, μετὰ δύο πού μῆνας εἰσεκομισάμην αὐτῷ μαλακοὺς ἀκρέμονας θάμνων τε καὶ φυτῶν, ὧν πάλιν καὶ αὐτῶν ὀσμησάμενον ἁπάντων, ἐνίων μὲν εὐθέως ἀπέστη, τινῶν δὲ ἠξίωσε γεύσασθαι, καὶ γευσάμενον ἐπὶ τὴν ἐδωδὴν ἐτράπετο τῶν καὶ ταῖς μεγάλαις αἰξὶ συνήθων ἐδεσμάτων. Ἀλλὰ τοῦτο μὲν ἴσως μικρόν· ἐκεῖνο δὲ μέγα. Τὰ γὰρ φύλλα καὶ τοὺς μαλακοὺς ἀκρέμονας ἀποφαγὼν κατέπιεν· εἶτ' ὀλίγον ὕστερον ἐπὶ τὸ μηρυκάζειν ἧκεν, ὃ πάλιν θεασάμενοι πάντες ἀνεβόησαν ἐκπλαγέντες ἐπὶ ταῖς τῶν ζῴων δυνάμεσι· μετὰ μὲν γὰρ ἢ καὶ τὸ πεινῆσαι διά τε τοῦ στόματος καὶ τῶν ὀδόντων προφέρεσθαι τὴν ἐδωδήν· ἀλλ' ὅτε τὸ καταποθὲν εἰς τὴν γαστέρα πρῶτον μὲν ἀναφέρειν εἰς τὸ στόμα προσῆκεν, ἔπειτα λεαίνειν ἐν αὐτῷ μασσώμενον ἐν χρόνῳ πόλλῳ, καὶ μετὰ ταῦτα καταπίνειν μηκέτι εἰς τὴν αὐτὴν κοιλίαν, ἀλλ' εἰς ἑτέραν, ἱκανῶς ἡμῖν ἐφαίνετο θαυμάσιον εἶναι. Παρορῶσι δὲ πολλοὶ τὰ τοιαῦτα τῆς φύσεως ἔργα, μόνα τὰ ξένα θεάματα θαυμάζοντες.

That is to say:

'Nature forming, fashioning, and perfecting the parts of the body, hath so brought it to pass, that they should of themselves without any teaching set about and perform their proper actions: and of this I once made a great experiment, bringing up a kid without ever seeing its dam. For dissecting some goats great with young, to resolve some questions made by anatomists, concerning the economy of nature in the formation of the fœtus in the womb ; and finding a brisk embryon (young one) I loosed it from the matrix after our usual manner, and snatching it away, before it saw its dam, I brought it into a certain room ; having many vessels full, some of wine, some of oil, some of honey, some of milk, or some other liquor ; and others not a few filled with all sorts of grain, as also with several fruits, and there laid it. This embryon we saw first of all getting up on its feet and walking, as if it had heard, that its legs were given it for that purpose ; next shaking off the slime it was besmeared with from the womb ; and moreover thirdly, scratching its side with one of its feet ; then we saw it smelling to every one of those things that were set in the room, and when it had smelt to them all, it supped up the milk ; whereupon we all for admiration cried out, seeing clearly the truth of what Hippocrates saith, That the natures and actions of animals are not taught, (but by instinct.) Hereupon I nourished and reared this kid, and observed it afterward not only to eat milk, but some other things that stood by it. And the time when this kid was taken out of the womb being about the vernal equinox, after some two months were brought into it the tender sprouts of shrubs and plants, and it again smelling of all of them instantly refused some, but was pleased to taste others ; and after it had tasted began to eat of such as are the usual food of goats. Perchance this may seem a small thing, but what I shall now relate is great. For eating the leaves and tender sprouts, it swallowed them down, and then a while after it began to chew the cud. Which all that saw cried out again with admiration, being astonished at the instincts and natural faculties of animals. For it was a great thing that when the creature was hungry it should take in the food by the mouth and chew it with its teeth; but that it should bring up again into the mouth that which it had swallowed down into its first stomach; and chewing it there a long time it should grind and smooth it, and afterward swallow it again, not into the same stomach, but into another, seemed to us wonderful indeed. But many neglect such works of nature, admiring only strange and unusual sights.' So far Galen.

This pleasant and admirable story, should one consider all the particulars of it, and endeavour to give an account of them, as also the inferences that might be drawn from it, one might fill a whole volume with comments upon it. All that I shall at present say is this, that in all this economy and these actions, counsel and design doth so clearly appear, that he must needs be very stupid that doth not discern it, or impudent that can deny it. I might add, that there seems to be something more than can be performed by mere mechanism in the election this creature made of its food: for before it would eat of any, it smelt to all the liquors before it, and when it had done so, betook itself to the milk and devoured that. He doth not say that the milk was the last liquor it smelt to, or that when it had once smelt to that it presently drank it up. The like also he saith of all the sprouts and branches of plants that were laid before it. By the bye, we may take notice of one thing very remarkable, that this kid of its own accord drank milk, after the manner it had done in the womb; whereas had it once drawn by the nipple, it would hardly have supped the milk. And therefore in weaning young creatures the best way is never to let them suck the paps at all, for then they will drink up milk without any difficulty; whereas if they have sucked, some will very hardly, others by no means be brought to drink. But how do the young with such facility come to take the nipple and to suck at it, which they had never before used to do? Here we must have recourse to natural instinct, and the direction of some superior cause.

Notice hath been already taken in an observation communicated by my learned friend Dr. Tancred Robinson of the providence of nature in so forming the membranes of the body as to be capable of a prodigious dilatation and extension; which is of so great use in some diseases, for example the dropsy, to continue life for some time till remedy may be had, and if not, to give time to prepare for death. But the wisdom and design of this texture doth in no instance more clearly appear than in the necessity of it for the womb in the time of gestation. For were not the womb in women, which during virginity is not bigger than a small purse, almost infinitely dilatable, and also the peritoneum, not to mention the skin and the cuticula; how were it possible it should contain the child, nay sometimes twins, with all their apurtenances, the secundines, the placenta, the liquors or waters, and whatever else is necessary for the defence, nutrition, respiration, and soft and convenient lodging of them, till they come to their due perfection and maturity for exclusion? How could the child have room to grow there to his bigness, and stir and turn himself as is requisite? Add here to another observation of Blasius's particularly relating to this subject. He hath observed that the vessels of the interior glandulous substance of the womb are strangely contorted and reflected with turnings and meanders, that they might not be too much strained, but their folds being extended and abolished, they might accommodate themselves without danger of rupture to the necessary extension of the uterine substance at that time.

Another remarkable proof of counsel and design may be fetched from the formation of the veins and arteries near the heart, which I meet with in Dr. Lower's treatise De Corde. Just before the entrance of the right auricle of the heart (saith he), to wit, where the ascending trunk of the vena cava meeting with the descending, is ready to empty itself into the said right auricle, there occurs in it a very remarkable knob or bunch [tuberculum] raised up from the subjacent fat; by the interposition whereof the blood falling down by the descending vein is diverted into the auricle, which otherwise encountring and bearing upon that of the ascendent trunk, would very much hinder and retard the motion of it upward towards the heart. And because in an erect site and figure of the body there is a greater and more eminent danger of such an accident, therefore the vena cava in mankind hath this tubercle far greater, and more extant than it is in brutes, so that if you thrust your finger into either trunk, you can hardly find passage or admittance into the other.

But in quadrupeds, as in sheep, dogs, horse, kine, in which the course of the blood from either extreme of the body is more equal, and as it were in a plainer level, and because the heart by reason of its bulk and weight hanging downwards, both trunks of the vena cava have some little declivity towards it, there is no need of so great a bar and diversion in them; yet are they not altogether devoid of it.

Moreover, lest the blood here in its conflux should make a kind of flood or whirl pool, whilst the auricle being contracted doth not give it free ingress, therefore in this place the vena cava in greater animals, as well man as quadrupeds, is round about musculous; as well that it may be restrained and kept within its due limits of extension, as that it may more vigorously and strongly urge and impel the blood into the cavity of the auricle.

Besides, there is no less providence and caution used, that the blood when it is forcibly cast out of the left ventricle of the heart, be not unequally distributed to the superior or inferior parts. For whereas this gate or orifice of the heart opens right upwards, if that channel which receives the first impulse of the blood did lead in a strait line up to the region of the head, it could not be, but that it must be poured too swiftly upon the brain, and so the inferior parts of the body must need be defrauded of their vital liquor and aliment. Which inconvenience that the divine Architect of the body might wholly obviate and avoid, in animals whose hearts are more strongly moved, he so artificially contrived the trunk of the aorta, which is next the heart, that the blood runs not directly into the axillary and carotide arteries, but doth, as it were, fetch a compass: for in the middle space between the ventricle and those arteries, it is very much inflected or bent; whence it comes to pass, that that crooked angle sustains the force and first stroke of the ejected blood, and directs the greatest torrent of it towards the descending trunk of the aorta, which otherwise would rush too forcibly into the superior branches thereof, distending immoderately, and soon oppress and burthen the head. So far Dr. Lower.

To elude or evade the force of all these instances,

and innumerable others which might be produced to demonstrate that the bodies of men and all other animals were the effects of the wisdom and power of an intelligent and almighty Agent, and the several parts and members of them designed to the uses to which now they serve, the atheist had one subterfuge, in which he most confides, viz. That all these uses of parts are no more than what is necessary to the very existence of the things to whom they belong : and that things made uses, and not uses things.

> ———Nil ideo natum est in corpore ut uti
> Possemus, sed quod natum est id procreat usum.
> *Lucretius, Lib.* 4.

And having instanced in several members, he concludes,

> ———Omnia denique; membra
> Ante fuere, ut opinor, eorum quamfuit usus.

I shall give you their sense together with the confutation of it in Dr. Bentley's words, borrowed, out of his fifth lecture, &c. and superadd something of my own.

These things (say they*) are mistaken for tokens of skill and contrivance, whereas they are but necessary consequences of the present existence of those creatures to which they belong. For he that supposeth any animals to subsist, doth by that very supposition allow them every member and faculty that are necessary to subsistence. And therefore, unless we can prove *à priori*, and independent on this usefulness, now that things are once supposed to have existed and propagated; that among almost infinite trials and essays at the beginning of things, among millions of monstrous shapes and imperfect formations, a few such animals as now exist could not possibly be produced, these after-considerations are of very little moment : because, if such ani-

* The atheists.

IN THE CREATION.

mals could in that way possibly be formed, as might live and move and propagate their beings, all this admired and applauded usefulness of their several fabrics is but a necessary condition and consequence of their existence and propagation.

This is the last pretence and sophistry of the atheists against the proposition in my text (Acts xvii. 27.) That we received our life and being from a divine wisdom and power. And as they cannot justly accuse me of concealing or baulking their grand objection; so I believe these following considerations will give them no reason to boast, that it cannot receive a just and satisfactory answer.

(1.) First therefore, we affirm that we can prove, and have done it already by arguments *à priori* (which is the challenge of the atheists), that these animals that now exist, could not possibly have been formed at first by millions of trials. For seeing they allow by their very hypothesis (and without standing to that courtesy, we have proved it before) that there can be no casual or spontaneous motion of the particles of matter, it will follow that every single monster among so many supposed myriads, must have been mechanically and necessarily formed according to the known laws of motion, and the temperament and quality of the matter it was made of. Which is sufficient, that no such monsters were or could have been formed. For to denominate them even monsters, they must have had some rude kind of organical bodies, some stamina of life, though never so clumsy, some system of parts, compounded of solids and liquids, that executed (though but bunglingly) their peculiar motions and functions. But we have lately shewn it impossible for nature unassisted to constitute such bodies, whose structure is against the law of specific gravity. So that she could not make the least endeavour towards the producing of a monster, or of any thing that hath more vital and organical parts than we find in a rock of marble, or a fountain of water. And again, though we should not contend with them about their monsters and abortions, yet seeing that they suppose even the perfect animals, that are still in being, to have been formed mechanically among the rest, and only add some millions of monsters to the reckoning, they are liable to all the difficulties in the former explication, and are expressly refuted through the whole preceding sermon, where it is abundantly shewn, that a spontaneous production is against the catholic laws of motion, and against matter of fact, a thing without example, not only in man and the nobler animals, but in the smallest of insects and the vilest of weeds: though the fertillity of the earth cannot be said to have been impaired since the beginning of the world.

(2.) Secondly, we may observe, that this evasion of the atheist is fitted only to elude such arguments of divine wisdom as are taken from things necessary to the conservation of the animal, as the faculties of sight, and motion, and nutrition, and the like; because such usefulness is indeed included in a general supposition of the existence of that animal, but it miserably fails him against other reasons from such members and powers of the body, as are not necessary absolutely to living and propagating, but only much conduce to our better subsistence and happier condition. So the most obvious contemplation of the frame of our bodies, as that we all have double sensories, two eyes, two ears, two nostrils, is an effectual confutation of this atheistical sophism. For a double organ of these senses is not at all comprehended in the notion of bare existence, one of them being sufficient to have preserved life, and continued the species, as common experience witnesseth. Nay even the very nails of our fingers are an infallible token of design and contrivance; for they are useful and convenient to give strength and firmness to those parts in the various functions they are put to; and to defend the numerous nerves and tendons that are under them, which have a most exquisite sense of pain, and without that native armour would continually be exposed to it. It is manifest therefore that there was a contrivance and foresight of the usefulness of nails antecedent to their formation. For the old stale pretence of the atheists, that things were first made fortuitously, and afterward their usefulness was observed or discovered, can have no place here; unless nails were either absolutely requisite to the existence of mankind, or were found only in some individuals, or some nations of men, and so might be ascribed to necessity upon one account, or to fortune upon another. But from the atheists' supposition, that among the infinite diversity of the first terrestial productions, there were animals of all imaginable shapes and structures of body, all of which survived and multiplied, that by reason of their make and fabric could possibly do so, it necessarily follows that we should now have some nations without nails upon their fingers, others with one eye only, as the poets describes the Cylopes in Sicily, and the Arimaspi in Scythia; others with one ear, or with one nostril, or indeed without any organ of smelling, because that sense is not necessary to man's subsistence; others destitute of the use of language, seeing that mutes also may live. One people would have the feet of goats, as the feigned Satyrs and Panisci: another would resemble the heads of Jupiter Ammon, or the horned statues of Bacchus; the Sciapodes and Enotocetæ, and other monstrous nations, would be no longer fables, but real instances in nature: and in a word, all the ridiculous and extravagant shapes that can be imagined, all the fancies and whimsies of poets and painters, and Egyptian idolaters, if so be they are consistent with life and propagation, would be now actually in being, if our atheists' notion were true: which therefore may deservedly pass for a mere dream and an error, till they please to make new discoveries in *terra incognita*, and bring along with them some savages of all these fabulous and monstrous configurations. Thus far Dr. Bentley, who adds four considerations more to confute this fancy, *ex abundanti*, granting the atheist all the absurd suppositions he can make. For which, though they be very well worth the reading, yet being too long to transcribe, I refer the reader to the sermon itself.

I shall now farther prove by a notable instance, that uses made things, that is to say, that some things were made designedly and on purpose for such a use as they serve to; and that is, the tendrels or claspers of plants, because they are given only to such species as have weak and infirm stalks, and cannot raise up or support themselves by their own strength. We see not so much as one tree, or shrub, or herb, that hath a firm and strong stem, and that is able to mount up and stand alone without assistance, furnished with them. Whereas had they been without design scattered (as I may say) indifferently and carelessly among plants, it could not possibly have happened, but among so many thousand species, they must have fallen to the lot of some few, at least some one of the strong, and not only of the weak. The same hath been proved by the instance of the power given to the hedgehog and armadillo, of contracting their bodies into a globular figure, and so hiding and securing their tender and unarmed parts.

2. I shall prove by another eminent instance, that things did not make uses, because there is a sort of creatures which have all the parts and organs which are fitted for a certain action, and employed for the exercise of it by another sort; and yet make no use of them for that purpose. That is, the ape-kind. The Parisian academists, in their anatomy of some animals of this kind, tell us, that the mucles of the os hyoïdes, tongue, larynx, and pharynx, which do most serve to articulate a word, were wholly like to those of man, and a great deal more than those of the hand; which nevertheless the ape, which speaks not, uses with as much perfection as a man. Which demonstrates that speech is an action more peculiar to man, and which more distinguishes him from brutes, than the hand, which Anaxagoras, Aristotle, and Galen have thought to be the organ which nature has given to man, as to the wisest of all animals; for want perhaps of this reflection. For the ape is found provided by nature of all those marvellous organs of speech with so much exactness, that the very three small muscles, which do take their rise from the apophysis styloïdes, are not wanting, although this apophysis be extremely small. This particularity does likewise shew, that there is no reason to think, that agents do perform such and such actions, because they are found with organs proper thereunto: for, according to these philosophers, apes should speak, seeing that they

have all the instruments necessary for speech. All this is confirmed and approved by the learned and accurate Dr. Tyson, in his anatomy of the orang-outang, or pigmy, he finding in the animal he described (which was of the ape-kind) the whole structure of the larynx and os hyoïdes exactly as it is in man. And the reflection which the Parisians make upon their observation of these and the neighbouring parts, he thinks very just and valuable; and adds farther, that this is not the only instance, which may justify such an inference, though he thinks it so strong a one, as the atheist can never answer.

It is farther considerable, and adds to the weight of this instance, that though birds have been taught to imitate human voice, and to pronounce words, yea sentences, yet quadrupeds never; though they have organs far more fit for that purpose, and some of them, *viz.* dogs, and horses, converse almost perpetually with men; and others, as apes, are given naturally to imitate men's actions; as if providence had designed purposely to confute this fond conceit of the atheists, by denying them the power to make use of these organs of speech, which whether they understand what they said or not, they otherwise might and would have done in imitation of man, and that to greater perfection than birds do or are capable of doing.

Farther to prove, that those nobler faculties of the soul, reason, and understanding, cannot be produced by matter organized, but must have a higher principle, he thus argues,* it is an observation of Vesalius's, that the brain of man, in respect of his body, is much larger than what is to be met with in any other animals, exceeding in bigness three oxen's brains; whence he infers, that as animals excel in the largeness of the brain, so they do likewise in the principal faculties of the soul, which inference the Doctor cannot allow.

It is (saith he) a generally received opinion, that the brain is the immediate seat of the soul itself, whence one would be apt to think, that seeing there is so great a disparity between the soul of a man and a brute, the organ in which it is placed should be very different too; yet by comparing the brain of our pigmy (the orangoutang, or wild-man) with that of a man, and with the greatest exactness observing each part in both, it was very surprising to me to find so great a resemblance of the one to the other as nothing could be more; and that in proportion to its body, its brain was also as large as a man's.

Since therefore (he proceeds) the brain of our pigmy does in all respects so exactly resemble a man's, I might here make the same reflection the Parisians did upon the organs of speech, that there is no reason to think, that agents do perform such and such actions, because they are found with organs proper thereunto; for then our pigmy might be really a man. The organs in animal bodies are only a regular compages of pipes and vessels for the fluid to pass through, and are passive. What actuates them are the humours and fluids: and animal life consists in their due and regular motion in this organical body. But those nobler faculties in the mind of man must certainly have a higher principle, and matter organized could never produce them: for why else, where the organ is the same, should not the action be the same too?

Object. Some may here object and argue, If the body of man be thus perfect, why did God make any other animals? for the most perfect being the best, an infinitely good agent, which wants neither wisdom nor power, should (one would think) only produce the most perfect.

Answ. To which I answer, 1. That according to this argumentation one might infer, that God must produce but one kind of creature, and that the most perfect that he is able, which is imimpossible: for he being infinite in all perfections, cannot act *ad extremum virium*, unless he could produce an infinite creature, that is, another God, which is a contradiction: but whatever he makes, must want degrees of infinite perfection, of which he could still (if he pleased) add more and more to it.

2. The inferior creatures are perfect in their order and degree, wanting no quality or perfection that is necessary or due to their nature and condition, their place and manner of living. Now, why God might not make several subordinate ranks and degrees of creatures, they being all good, I see no reason.

3. These several ranks and degrees of creatures are subservient one to another; and the most of them serviceable, and all some way or other useful to man; so that he could not well have been without them.

4. God made these several orders and degrees, and in each degree so many varieties of creatures, for the manifestation and displaying of his infinite power and wisdom. For we have shewn before by a familiar instance, that there is more art and wisdom shewn in contriving and forming a multitude of differing kinds of engines, than in one only.

5. Yet do I not think, that he made all these creatures to no other end, but to be serviceable to man, but also to partake themselves of his overflowing goodness, and to enjoy their own

beings. If we admit all other creatures in this inferior world, besides man, to be mere machines or *automata*, to have no life, no sense or perception of any thing, then I confess this reason is out of doors ; for being uncapable of pleasure or pain, they can have no enjoyment. Upon this account also, among others, I am less inclinable to that opinion.

I should now proceed to answer some objections which might be made against the wisdom and goodness of God in the contrivance and governance of the world, and all creatures therein contained. But that is too great and difficult a task for my weakness, and would take up more time than I have at present to spare, were I qualified for it, and besides swell this volume to too great a bulk. Only I shall say something to one particular, which was suggested to me by a learned and pious friend.*

Object. A wise agent acts for ends. Now what end can there be of creating such a vast multitude of insects as the world is filled with ; most of which seem to be useless, and some also noxious and pernicious to man, and other creatures?

Answ. To this I shall answer, 1. As to the multitude of species or kinds. 2. As to the number of individuals in each kind.

First as to the multitude of species (which we must needs acknowledge to be exceeding great, they being not fewer, perchance more, than 20,000). I answer, there were so many made,

1. To manifest and display the riches of the power and wisdom of God, Psa. civ. 24. 'The earth is full of thy riches : so is this great and wide sea; wherein are things creeping innumerable,' &c. We should be apt to think too meanly of

* Mr. Robert Burscough, of Totness in Devon.

those attributes of our Creator, should we be able to come to an end of all his works, even in this sublunary world. And therefore I believe never any man yet did, never any man shall, so long as the world endures, by his utmost industry attain to the knowledge of all the species of nature. Hitherto we have been so far from it, that in vegetables, the number of those which have been discovered this last age hath far exceeded that of all those which were known before. So true is that we quoted before out of Seneca, 'Pusilla res est mundus, nisi in eo quod quarat omnis mundus habeat.' The world is so richly furnished and provided, that man need not fear want of employment, should he live to the age of Methuselah, or ten times as long. But of this, having touched it already, I shall add no more.

2. Another reason why so many kinds of creatures were made, might be to exercise the contemplative faculty of man; which is in nothing so much pleased as in variety of objects. We grow weary of one study; and if all the objects of the world could be comprehended by us, we should, with Alexander, think the world too little for us, and grow weary of running in a round of seeing the same things. New objects afford us great delight, especially if found out by our own industry. I remember Clusius saith of himself, that upon the discovery of a new plant, he did not less rejoice than if he had found a rich treasure. Thus God is pleased, by reserving things to be found out by our pains and industry, to provide us employment most delightful and agreeable to our natures and inclinations.

3. Many of these creatures may be useful to us, whose uses are not yet discovered, but reserved for the generations to come, as the uses of some we now know are but of late invention, and were

IN THE CREATION. 305

unknown to our forefathers. And this must needs be so, because, as I said before, the world is too great for any man or generation of men, by his or their utmost endeavours, to discover and find out all its store and furniture, all its riches and treasures

Secondly, As to the multitude of individuals in each kind of insect, I answer,

1. It is designed to secure the continuance and perpetuity of the several species; which if they did not multiply exceedingly, scarce any of them could escape the ravine of so many enemies as continually assault and prey upon them, but would endanger to be quite destroyed and lost out of the world.

2. This vast multitude of insects is useful to mankind, if not immediately, yet mediately. It cannot be denied that birds are of great use to us; their flesh affording us a good part of our food, and that the most delicate too, and their other parts physic, not excepting their very excrements. Their feathers serve to stuff our beds and pillows, yielding us soft and warm lodging, which is no small convenience and comfort to us, especially in these northern parts of the world. Some of them have also been always employed by military men in plumes, to adorn their crests, and render them formidable to their enemies. Their wings and quills are made use of for writing pens, and to brush and cleanse our rooms and their furniture. Besides, by their melodious accents they gratify our ears; by their beautiful shapes and colours; they delight our eyes, being very ornamental to the world, and rendering the country where the hedges and woods are full of them, very pleasant and cheerly, which without them would be no less lonely and melancholy. Not to mention the exercise, diversion, and recreation which some of them give us.

Now insects supply land-birds the chiefest part

of their sustenance; some, as the entire genus of swallows, living wholly upon them, as I could easily make out, did any man deny or doubt of it: and not swallows alone, but also woodpeckers, if not wholly, yet chiefly; and all other sorts of birds partly, especially in winter-time, when insects are their main support, as appears by dissecting their stomachs.

As for young birds, which are brought up in the nest by the old, they are fed chiefly, if not solely, by insects. And therefore the time when birds for the most part breed is the spring, when there are multitudes of caterpillars to be found on all trees and hedges. Moreover it is very remarkable, that of many such birds, as when grown up, feed almost wholly upon grain, the young ones are nourished by insects. For example, pheasants and partridges, which are well known to be granivorous birds, the young live only or mostly upon ants' eggs. Now birds being of a hot nature, are very voracious creatures, and eat abundantly, and therefore there had need be an infinite number of insects produced for their sustenance. Neither do birds alone, but many sorts of fishes feed upon insects, as is well known to anglers, who bait their hooks with them. Nay, which is more strange, divers quadrupeds feed upon insects, and some live wholly upon them, as two sorts of tamunduus upon ants, which therefore are called in English ant-bears: the chameleon upon flies; the mole upon earth-worms. The badger also lives chiefly upon beetles, worms, and other insects.

Here we may take notice by the way, that because so many creatures live upon ants and their eggs, providence hath so ordered it, that they should be the most numerous of any tribe of insects that we know.

Conformable to this particular is the reason

IN THE CREATION. 307

my ingenious friend Mr. Derham, before remembered, hath given of the production of such innumerable multitudes of some aquatic insects.

I have often thought (saith he) that there was some more than ordinary use in the creation for such insects as are vastly numerous. Such as the *pulices aquatici*, which are in such swarms as to discolour the waters, and many others: and therefore I have bent my inquiries to find out the uses of such creatures; wherein I have so far succeeded as to discover, that those vastly small animalcula, not to be seen without a microscope, with which the waters are replete, serve for food to some others of the small insects of the waters, particularly to the *nympha culicaria*, (*hirsuta* it may be called), figured in Swammerdam. For viewing that nympha one day, to observe the motion of its mouth, and for what purpose it is in such continual motion; whether as fish to get air, or to suck in food, or both, I could plainly perceive the creature to suck in many of these most minute animalcula, that were swimming briskly about in the water. Neither yet do these animalcules serve only for food to such nymphæ, but also to another, to me anonymous, insect of the waters, of a dark colour, cleft as it were in sunder, and scarce so big as the smallest pin's head. These insects hunt these animalcules, and other small creatures that occur in the water, and devour them: and I am apt to think, although I have not yet seen it, that the *pulex aquaticus arborescens* liveth upon these or more minute and tender animalcules, and that it is to catch them that it so leaps in water.

This to me seems a wonderful work of God, to provide for the minutest creatures of the waters food proper for them, that is, minute and tender, and fit for their organs of swallowing.

As for noxious insects, why there should be so many of them produced, if it be demanded,

I answer, 1. That many that are noxious to us, are salutary to other creatures; and some that are poison to us, are food to them. So we see the poultry-kind feed upon spiders. Nay, there is scarce any noxious insect but one bird or other eats it, either for food or physic. For many, nay most of those creatures whose bite or sting is poisonous, may safely be taken entire into the stomach. And therefore it is no wonder that not only the ibis of Egypt, but even storks and peacocks prey upon and destroy all sorts of serpents as well as locusts and caterpillars.

2. Some of the most venomous and pernicious of insects afford us noble medicines, as scorpions, spiders, and cantharides.

3. These insects seldom make use of their offensive weapons, unless assaulted or provoked in their own defence, or to revenge an injury. Let them but alone and annoy them not, nor disturb their young, and unless accidentally, you shall seldom suffer by them.

Lastly, God is pleased sometimes to make use of them as scourges, to chastise or punish wicked persons or nations, as he did Herod and the Egyptians. No creature so mean and contemptible, but God can, when he pleases, produce such armies of them as no human force is able to conquer or destroy, but they shall of a sudden consume and devour up all the fruits of the earth, and whatever might serve for the sustenance of man, as locusts have often been observed for to do.

Did these creatures serve for no other use, as they do many; yet those that make them an objection against the wisdom of God, may (as Dr. Cockburn well notes) as well upbraid the prudence and policy of a state for keeping forces, which generally are made up of very rude and insolent people, which yet are necessary either to suppress rebellions, or punish rebels and other disorderly and vicious persons, and keep the world in quiet.

From that part of this discourse which relates to the body of man, I shall make these practical inferences.

Infer. 1. First, let us give thanks to Almighty God for the perfection and integrity of our bodies. It would not be amiss to put it into the eucharistical part of our daily devotions: we praise thee, O God, for the due number, shape and use of our limbs and senses, and in general of all the parts of our bodies; we bless thee for the sound and healthful constitution of them: Psal. x. 'It is thou that hast made us and not we ourselves; in thy book were all our members written.' The formation of the body is the work of God; and the whole process thereof attributed to him, Psal. cxxix. 13, 14, 15. The mother that bears the child in her womb is not conscious to any thing that is done there; she understands no more how the infant is formed, than itself doth. But if God hath bestowed upon us any peculiar gift or endowment, wherein we excel others, as strength, or beauty, or activity, we ought to give him special thanks for it, but not to think the better of ourselves, therefore, or despise them that want it.

Now because these bodily perfections, being common blessings, we are apt not at all to consider them, or not to set a just value on them; and because the worth of things is best discerned by their want, it would be useful sometimes to imagine or suppose ourselves by some accident to be deprived of one of our limbs or senses, as a hand, or a foot, or an eye, for then we cannot but be sensible, that we should be in worse condition than now we are, and that we should soon find a difference between two hands and one hand, two eyes and one eye, and that two excel one as much in worth as they do in number; and yet if we could spare the use of the lost part, the deformity and unsightliness of such a defect in the body, would alone be very grievous to us. Again, which is less, suppose we only, that our bodies want of their just magnitude, or that they or any of our members are crooked or distorted, or disproportionate to the rest either in excess or defect; nay, which is least of all, that the due motion of any one part be perverted, as but of eyes in squinting, the eye-lids in twinkling, the tongue in stammering, these things are such blemishes and offences to us, by making us gazing-stocks to others, and objects of their scorn and derision, that we could be content to part with a good part of our estates to repair such defects, or heal such infirmities. These things considered and duly weighed, would surely be a great and effectual motive to excite in us gratitude for this integrity of our bodies, and to esteem it no small blessing, I say a blessing and favour of God to us; for some there be that want it, and why might not we have been of that number? God was no way obliged to bestow it upon us.

And as we are to give thanks for the integrity of our body, so are we likewise for the health of it, and the sound temper and constitution of all its parts and humours; health being the principal blessing of this life, without which we cannot enjoy, or take comfort in any thing besides.

Neither are we to give thanks alone for the first collation of these benefits, but also for their preservation and continuance. God preserves our souls in life, and defends us from dangers and sad accidents, which do so beset us on every side, that the greatest circumspection in the world could not secure us, did not his good providence continually watch over us. We may be said to walk and converse in the midst of snares; besides, did we but duly consider the make and frame of our bodies, what a multitude of minute parts and vessels there are in them, and how an obstruction in one redounds to the prejudice of the whole, we could not but wonder how so curious an engine as man's body could be kept in tune one hour, as we use it, much less hold out so many years: how it were possible it should endure such hardship, such blows, so many shocks and concussions, nay such violences and outrages as are offered it by our frequent excesses, and not be disordered and rendered useless; and acknowledge the transcendent art and skill of Him who so put it together, as to render it thus firm and durable.

Infer. 2. Secondly, Have a care thou dost not by any vicious practice deface, mar, or destroy the workmanship of God. So use this body as to preserve the form and comeliness, the health and vigour of it.

1. For the form and beauty of the body, which mankind generally is fond enough of: and which must be acknowledged to be a natural endowment and blessing of God, a thing desirable, which all men take complacency in; which renders persons gracious and acceptable in the eyes of others; of which yet we do not observe, that brute beasts take any notice at all: of this I shall observe, that outward beauty is a sign of inward; and that handsome persons are naturally well inclined, till they do either debauch themselves, or are corrupted by others; and then with their

manners they mar their beauty too. For a man may observe and easily discern, that as persons are better or worse inclined, the very air of their visage will alter much; and that vicious courses, 'defacing the inward pulchritude of the soul, do change even the outward countenance into an abhorred hue:'* as is evident in the vices of intemperance and anger, and may by sagacious persons be observed in others also. No better cosmetics than a severe temperance and purity, a real and unaffected modesty and humility, a gracious temper and calmness of spirit, a sincere and universal charity. No true beauty without the signatures of these graces in the very countenance. They therefore who through the contrary vices do deface and blot out this natural character and impress, and do violence to their own inclinations, that sacrifice this jewel to their lusts, that reject this gift of God, and undervalue the favour of man, aggravate their sin and misery, and purchase hell at somewhat a dearer rate than others do. And those that have but a mean portion of this gift, are the more obliged by virtuous practice, not only to preserve, but to improve it. Virtue (as Cicero observes) if it could be seen with corporeal eyes, 'admirabiles sui amores excitaret.' 'It would excite a wonderful love of itself.' By the signatures it there impresses, it is in some measure visible in the faces of those that practise it, and so must needs impart a beauty and amiableness to them.

Diogenes Laertius in the life of Socrates tells us, that that philosopher was wont to advise young men συνεχὲς κατοπτρίζεσθαι, often to behold themselves in their looking-glasses or mirrors. Grammercy Socrates, that is good counsel indeed, will our young gentlemen and ladies be ready to

* Dr. More.

IN THE CREATION. 313

say, we like it very well, and we practise accordingly; and it seems we are injuriously taxed and reprehended by divines, for spending so much time between a comb and a glass. Be not overhasty, take what remains along with you: mark the end for which the philosopher exhorts this, ἵν᾿ εἰ μὲν καλοὶ εἶεν, ἄξιοι γίγνοιντο, εἰ δ᾿ αἰσχροὶ παιδείᾳ τὴν δυσείδειαν ἐπικαλύπτοιεν. 'That if they be handsome, they might approve themselves worthy of their form; but if they be otherwise, they may by discipline and institution hide their deformity.' And so by their virtuous behaviour compensate the hardness of their favour, and by the pulchritude of their souls, make up what is wanting in the beauty of their bodies. And truly, I believe, a virtuous soul hath influence upon its vehicle, and adds a lustre even to the outward man, shining forth in the very face.

2. So use the body, as to preserve the health and vigour, and consequently produce the life of it. These are things that all men covet. No more effectual means for the maintenance and preservation of them, than a regular and virtuous life. That health is impaired by vice, daily experience sufficiently evinceth. I need not spend time to prove, what no man doth or can deny. And as for length of days, we find by the same experience, that intemperate and disorderly persons are for the most part short-lived: moreover, immoderate cares and anxiety are observed suddenly to bring grey hairs upon men, which are usually the signs and forerunners of death. And, therefore, the way to live long, must needs be in all points to use our bodies, so as is most agreeable to the rules of temperance, and purity, and right reason. Every violence offered to it weakens and impairs it, and renders it less durable and lasting. One means there is, which physi-

cians take notice of, as very effectual for the preservation of health, which I cannot here omit, that is, a quiet and cheerful mind, not afflicted with violent passions or distracted with immoderate cares; for these have a great and ill influence upon the body. Now how a man can have a quiet and cheerful mind under a great burden and load of guilt, I know not, unless he be very ignorant, or have a seared conscience. It concerns us therefore, even upon this account, to be careful of our conversations, and to keep our consciences void of offence both toward God, and toward men.

Infer. 3. Thirdly, Did God make the body, let him have the service of it. Rom. xii. 1. 'I beseech you, brethren, by the mercies of God, that you present your bodies a living sacrifice, holy, acceptable unto God, which is your reasonable service.' How we should do that, St. Chrysostom tells us in his commentary upon this place, Μηδὲν ὀφθαλμὸς πονηρὸν βλεπέτω, καὶ γέγονε θυσία· μηδὲν ἡ γλῶσσα λαλείτω αἰσχρὸν, καὶ γέγονε προσφορά· μηδὲν ἡ χεὶρ πραττέτω παράνομον, καὶ γέγονεν ὁλοκαύτωμα, &c. 'Let the eye behold no evil thing, and it is made a sacrifice; let the tongue speak no filthy word, and it becomes an oblation: let the hand do no unlawful action, and you render it a holocaust. Yet it is not enough thus to restrain them from evil; but they must also be employed and exercised in doing that which is good: the hand in giving alms, the tongue in blessing them that curse us and despitefully use us, the ear in hearkening to divine lectures and discourses.' 1 Cor. vi. 20. 'Glorify God in your body, or with your body, and in your spirits, which are God's,' and that not by redemption only, of which the apostle there speaks, but by creation also; Rom. vi. 12. 'Neither yield your members as instruments

IN THE CREATION. 315

of unrighteousness unto sin, but as instruments of righteousness unto God.' And again, ver. 19. 'Even so now yield your members servants of righteousness unto holiness.' I shall instance in two members, which are especially to be guarded and restrained from evil, and employed in the service of God.

First, The eye. We must 'turn away our eyes from beholding vanity,' as David prayed God would his, Psal. cxix. 37. We must 'make a covenant with our eyes,' as Job did, Job xxxi. 1. These are the windows that let in exterior objects to the soul: by these the heart is affected: this way sin entered first into the world. Our first parent saw that 'the tree' and its fruit 'was pleasant to the eyes,' and so was invited to take and eat it. There are four sins especially for which the eye is noted, as either discovering themselves in the eyes, or whose temptations enter in by, and so give denomination to the eye.

1. There is a 'proud eye,' Prov. xxx. 13. 'There is a generation, O how lofty are their eyes, and their eye-lids are lifted up.' Chap. vi. 17. A proud look is reckoned the first of those six things that God hates, Psal. xviii. 27. 'God' (the psalmist saith) 'will bring down' proud or 'high looks.' Psal. ci. 5. 'Him that hath a high look and proud heart' (saith David) 'I will not suffer.' And in Psalm cxxxi. 1. He saith of himself, that 'his heart is not haughty, nor his eyes lofty.' By which places it appeareth that pride sheweth forth itself in the eyes especially, and that they are as it were the seat or throne of it.

2. There is a 'wanton eye,' which the prophet Isaiah speaks of in his third chapter, at the 16th verse; 'Because the daughters of Jerusalem walk with stretched out necks, and wanton eyes.' The apostle Peter, in his second epistle, ii. 24. men-

tions 'eyes full of adultery.' For by these casements enter in such objects, as may provoke and stir up adulterous thoughts in the mind, as they did in David's; and likewise impure thoughts conceived in the heart may discover themselves by the motions of the eye. And, therefore, in this respect, we should do well with holy Job, to make a covenant with our eyes; not to gaze upon any object which may tempt us to any inordinate appetite or desire. For our Saviour tells us, it were better to pluck out our right eye, than that it should be an offence to us: which I suppose refers to this matter, because it immediately follows those words, 'He that looketh upon a woman to lust after her, hath already committed adultery with her in his heart.

3. There is a 'covetous eye.' By covetousness, I understand, not only a desiring what is another man's, which is forbidden in the tenth commandment, but also an inordinate desire of riches, which the apostle John seems to understand in his first epistle, ii. 16. by 'the lust of the eye.' And covetousness may well be called the 'lust of the eye,' because, 1. The temptation or tempting object enters by the eye. So the seeing the wedge of gold and Babylonish garment stirred up the covetous desire in Achan. 2. Because all the fruit a man reaps of riches more than will furnish his necessities and conveniences, is the feeding of his eye, or the pleasure he takes in the beholding of them, Eccles. v. 11. 'When goods increase, &c. what good is there to the owners thereof, saving the beholding them with their eyes?'

4. There is an 'envious eye,' which by our Saviour is called an evil eye; Matth. xx. 15. 'Is thine eye evil because I am good?' that is, enviest thou thy brother, because I am kind to him.

And, vii. 22. one of those evil things which proceed out of the heart and defile a man, is 'an evil eye.' Envy is a repining at the prosperity or good of another, or anger and displeasure at any good of another which we want, or any advantage another hath above us: as in the parable of the labourers in the vineyard, those that came in first envied the last, not because they received more than they, but because they received equal wages for less time. Those that are subject to this vice, cannot endure to see another man thrive; and are apt to think his condition better than theirs, when indeed it is not.

Let us then so govern our eyes, that we discover by them none of these vices. Let the humility and purity of our minds appear even in our outward looks. Let neither pride nor lust manifest themselves in the posture or motions of our eyes. Let us have a care that these members be neither the inlets, nor outlets of any of the forementioned vices; that they neither give admission to the temptation, nor be expressive of the conception of them. Let us employ them in reading the word of God, and other good books, for the increase of our knowledge, and direction of our practice; in diligently viewing and contemplating the works of the creation, that we may discern and admire the footsteps of the divine wisdom easily to be traced in the formation, disposition, and designations of them. Let us take notice of any extraordinary events and effects of God's providence towards ourselves or others, personal or national; that as they are the issues of his mercy or justice, they may stir up suitable affections in us, of thankfulness or fear. Let those sad and miserable objects, that present themselves to our sight, move us to pity and commiseration. And let our eyes sometimes be exercised in weeping for the miseries and calamities of others, but especially for our own and their sins.

Secondly, Another member I shall mention, is the tongue, which as it is the chief instrument of speech, so it may be well or ill employed in the exercise of that action, and therefore stands in need of direction and restraint. I remember I once heard from an ingenious anatomist of Padua this observation, that there are but two members in the body that have a natural bridle, both which do very much need it; the tongue, and another I shall not name. The signification whereof may be, that they are not to be let loose, but diligently curbed and held in. That the tongue needs a bridle, you will readily grant if you read what the apostle James hath written of it, chap. iii. 6. 'The tongue is a fire, a world of iniquity. So is the tongue among our members, that it defileth the whole body, and setteth on fire the course of nature, and is set on fire of hell. For every kind of beasts and of birds, and of serpents, and of things in the sea is tamed, and hath been tamed of mankind; but the tongue can no man tame, it is an unruly evil, full of deadly poison.' For the better government of the tongue, I shall note some vices of speech, which must carefully be avoided. First of all loquacity or garrulity. This the contrivance of our mouths suggests to us. Our tongues are fenced and guarded with a double wall or mound of lips and teeth, that our words might not rashly and unadvisedly slip out. Then nature hath furnished us with two ears, and but one tongue, to intimate, that we must hear twice so much as we speak. Why loquacity is to be avoided, the wise man gives us a sufficient reason, Prov. x. 19. 'In the multitude of words there wanteth not sin.' And

Eccles. v. 7. 'In many words there are divers vanities.' To which we may add another, of great force with most men, viz. That it hath been, always esteemed an effect and argument of folly, Eccles. v. 3. 'A fool's voice is known by multitude of words.' And on the contrary, to be of few words is a sign of wisdom; and he that is wise enough to be silent, though a fool, may pass undiscovered. Besides all this, a talkative person must needs be impertinent, and speak many idle words, and so render himself burdensome and odious to company; and may perchance run himself upon great inconveniences, by blabbing out his own or others' secrets; for a word once uttered, *fugit irrevocabile*, is irrevocable, whatever the consequence of it be. Great need therefore have we ' to set a watch over our mouths, and to keep the door of our lips;' Psal. cxli. 3. and not suffer our tongues προτρέχειν τῆς διανοίας;* as Isocrates phraseth it.

Secondly, Lying or falsely speaking. There is difference between *mentiri* and *mendacium dicere*, that is, lying and speaking of an untruth, or thing that is false. *Mentiri* is *contra mentem ire*, which though it be no good etymology of the word, is a good notion of the thing; that is, to go against one's mind, or speak what one does not think.

Ἕτερον μὲν κεύθειν ἐνὶ φρεσὶν, ἄλλο δὲ βάζειν,

As Homer expresses it, to conceal one thing in the mind, and speak another with the tongue. Hence a man may speak an untruth, and yet not lie, when he thinks he speaks the truth; and on the contrary, may speak what is materially true, and yet lie, when he speaks what he thinks not to be true. The tongue was made to be the index of the mind, speech the interpreter of

* Run before the understanding or wit.

thought; therefore there ought to be a perfect harmony and agreement between these two. So that lying is a great abuse of speech, and a perverting the very end of it, which was to communicate our thoughts one to another. It hath also an ill principle, for the most part proceeding either from baseness of spirit or cowardice, as in them that have committed a fault, and deny it, for fear of punishment or rebuke. And therefore the ancient Persians, as Xenophon tells us in his Κύρου παιδεία, made it one of the three things they diligently taught their children; which were ἱππεύειν, καὶ τοξεύειν, καὶ ἀληθεύειν, 'To ride, to shoot, and to speak the truth;' or from covetousness, as in tradesmen, who falsely commend their commodities, that they may vend them for a greater price; or from vanity and vain-glory, in them who falsely boast of any quality or action of their own. It is odious both to God and man. To God; Prov. vi. 17. ' a lying tongue' is one of those six or seven things that are an abomination to him. To men, as Homer witnesseth in the verse preceding the fore-quoted.

Ἐχθρὲς γάρ μοι κεῖνος ὁμῶς Ἀΐδαο πύλῃσι, &c.

'He that tells lies is as hateful to me as the gates of hell or death.'———The practice of lying is a diabolical exercise, and they that use it are the devil's children, as our Saviour tells us, John viii. 44. ' Ye are of your father the devil &c. for he is a liar, and the father of it.' And lastly, it is a sin that excludes out of heaven, and depresses the soul into hell, Rev. xxi. 8. 'All liars shall have their part in the lake which burns with fire and brimstone, which is the second death.'

Thirdly, Another vice or abuse of speech, or

vicious action to which the tongue is instrumental, is slandering; that is, raising a false report of any man tending to his defamation. This might have been comprehended under the former head, being but a kind of lying proceeding from enmity or ill will. It is a very great injury to our neighbour, men's reputation being as dear to them as life itself; so that it is grown to be a proverb among the vulgar, 'Take away my good name, and take away my life.' And that which enhances this injury, is that it is irreparable. We cannot by any contrary declaration so clear the innocency of our neighbour, as wholly to extirpate the pre-conceived opinion out of the minds of those to whom our confession comes; and many will remain whom the calumny hath reached, to whom the vindication probably will not extend; the pravity of man's nature being more apt to spread and divulge an ill report, than to stop and silence it. I might instance in flattering of others, and boasting of ourselves, for two abuses of speech, but they may both be referred to lying, the one to please others, and puff them up with self-conceit, and a false opinion that they have some excellent quality or endowment, which they want, or have not in such a degree, or that they are better thought of by others, than indeed they are, and more honoured. The other, to gain more honour than is due to ourselves. Neither yet is boasting only of what we have not, but also of what we have, condemned and disallowed by God and men, as being contrary to that humility and modesty that ought to be in us, Prov. xxvii. 2. 'Let another man praise thee and not thine own mouth; a stranger, and not thine own lips.' And moralists proceed so far as to censure all unnecessary περιαυτολογία, that is, talking of a man's self.

Fourthly, Obscene and impure words are another vicious effect of the tongue. Those are principally the σαπροὶ λόγοι, rotten speeches, the apostle speaks of, Eph. v. 29. Such as chaste ears abhor, which tend only to the depraving and corrupting the hearers; and are to be studiously and carefully avoided by all that pretend to Christianity; Eph. v. 3. 'But fornication and all uncleanness let it not be once named among you.'

Fifthly, Cursing and railing, or reviling words, are also a great abuse of speech, and outrageous effects and expressions of malice and wickedness; Psal. x. 7. The Psalmist makes it part of the character of a wicked man, that 'his mouth is full of cursing.' Which passage we have quoted by the apostle, Rom. iii. 14. 'Whose mouth is full of cursing and bitterness.'

Sixthly, Swearing and irreverent using the name of God in common discourse and converse, is another abuse of the tongue; to which I might add vehement asseverations upon slight and trivial occasions. I do not deny, but in a matter of weight and moment, which will bear out such attestation, and where belief will not be obtained without them, and yet it may much import the hearer or speaker that his words be believed, or where the hearer would not otherwise think the matter so momentous or important as indeed it is, protestations and asseverations, yea, oaths, may lawfully be used. But to call God to witness to an untruth or a lie perhaps, or to appeal to him on every trivial occasion, in common discourse, customarily, without any consideration of what we say, is one of the highest indignities and affronts that can be offered him, being a sin to which there is no temptation. For it is so far from gaining belief (which is the only thing that can with any show of reason be pleaded for it), that it rather creates diffidence and distrust. For as 'multa fidem promissa levant,' so ' multa juramenta' too, it being become a proverb, 'He that will swear will lie.' And good reason there is for it; for he that scruples not the breach of one of God's commands, is not likely to make conscience of the violation of another.

Lastly (for I will name no more), Scurrilous words, scoffing and jeering, flouting and taunting, are to be censured as vicious abuses of speech.

This scoffing and derision proceeds from contempt, and that of all injuries men do most impatiently bear; nothing offends more, or wounds deeper; and therefore what greater violation of that general rule of Christian practice 'to do to others as we would they should do unto us? This injury of being derided the Psalmist himself complains of, Psal. lxix. 11, 12. 'I became a proverb to them. They that sit in the gate speak against me, and I was the song of the drunkards.' And Psal. xxxv. 15. according to the church translation, 'The very abjects came together against me unawares, making mows at me, and ceased not;' and the prophet Jeremy, Jer. xx. 7. 'I am in derision daily, every one mocketh me.' And though there may be some wit shewn in scoffing and jesting upon others, yet it is a practice inconsistent with true wisdom. The scorner and the wise man are frequently opposed in Scripture. Prov. ix. 8. and xiii. 1, &c. It is a proverbial saying. 'The greatest clerks are not always the wisest men.' I think the saying might as often be verified of the greatest wits. Scorning in that gradation in the first Psalm is set down as the highest step of wickedness. And

Solomon tells us, 'That judgments are prepared for the scorners.'

You will say to me, how then must our tongues be employed? I answer, 1. In praises and thanksgiving unto God, Psal. xxxv. 28. 'And my tongue shall speak of thy righteousness and of thy praises all the day long.' Parallel whereto is ver. 24. of Psal. lxxi. Indeed the book of Psalms is in a great measure but an exercise of, or exhortation to, this duty. 2. We must exercise our tongues in talking of all his wondrous works; Psal. cxlv. 5, 6. 'I will speak of the glorious honour of thy majesty, and of thy wondrous works.' 3. In prayer to God. 4. In confession of him, and his religion, and publicly owning it before men, whatever the hazard be. 5. In teaching, instructing, and counselling of others. 6. In exhorting them. 7. In comforting them that need it. 8. In reproving them. All which particulars I might enlarge upon; but because they come in here only as they refer to the tongue, it may suffice to have mentioned them summarily.

Thirdly, Let us hence learn duly to prize and value our souls. Is the body such a rare piece, what then is the soul? The body is but the husk or shell, the soul is the kernel; the body is but the cask, the soul the precious liquor contained in it; the body is but the cabinet, the soul the jewel; the body is but the ship or vessel, the soul the pilot; the body is but the tabernacle, and a poor clay tabernacle or cottage too, the soul the inhabitant; the body is but the machine or engine, the soul that ἐνδόν τι, that actuates and quickens it; the body is but the dark lantern, the soul or spirit is the candle of the Lord that burns in it. And seeing there is such difference between the soul and the body in respect of excellency, surely our better part challenges our greatest care and diligence to make provision for it. Bodily provision is but half provision, it is but for one part of a man, and that the meaner and more ignoble too, if we consider only the time of this life; but if we consider a future estate of endless duration after this life, then bodily provision will appear to be, I do not say quarter provision, but no provision at all, in comparison; there being no proportion between so short a period of time, and the infinite ages of eternity. Let us not then be so foolish as to employ all our thoughts and bestow all our time and pains about cherishing, accommodating, and gratifying our bodies, in 'making provision for the flesh to fulfil the lusts thereof,' as the apostle phraseth it; and suffer our souls to lie neglected, in a miserable, and poor, and blind, and naked condition. Some philosophers will not allow the body to be an essential part of man, but only the vessel or vehicle of the soul; 'Anima cujusque est quisque.' 'The soul is the man.' Though I would not be so unequal to it, yet I must needs acknowledge it to be but an inferior part. It is therefore so to be treated, so dieted and provided, as to render it most calm and compliant with the soul, most tractable and obsequious to the dictates of reason; not so pampered and indulged, as to encourage it to cast its rider, and to take the reins into its own hand, and usurp dominion over the better part, the τὸ ἡγεμονικὸν, to sink and depress it into a sordid compliance with its own lusts, 'Atque affigere humi divinæ particulam auræ.'

This is our duty, but, alas, what is our practice? Our great partiality towards our bodies, and neglect of our souls, shews clearly which part we prefer. We are careful enough of wounding or maiming our bodies, but we make bold to lash and wound our souls daily; for every sin we commit, being contrary to its nature, is a real stripe, yea, a mortal wound to the soul, and we shall find it to be so, if our consciences be once awakened to feel the sting and smart of it. We are industrious enough to preserve our bodies from slavery and thraldom, but we make nothing of suffering our souls to be slaves and drudges to our lusts, and to live in the vilest bondage to the most degenerate of creatures, the devil. We are thrifty and provident enough not to part with any thing that may be serviceable to our bodies under a good consideration, and we so esteem them, as that we will part with all we have for the life of them; but we make little account of what is most beneficial to our souls, the means of grace and salvation, the word of God, and duties of his worship and service; nay, we can be content to sell our souls themselves for a trifle, for a thing of nothing, yea, for what is worse than nothing, the satisfying of an inordinate and unreasonable appetite or passion. We highly esteem and stand much upon our nobility, our birth and breeding, though we derive nothing from our ancestors but our bodies and corporeal qualities; and it is useful so far to value and improve this advantage, as to provoke us to imitate the good examples of our progenitors, not to degenerate from them, nor to do any thing unworthy of our breeding; and yet the divine original of our souls, which are beams from the Father of lights, and the immediate offspring of God himself, τοῦ γὰρ καὶ γένος ἐσμὲν, hath little influence upon us to engage us to walk worthily of our extraction, and to do nothing that is base or ignoble, and unsuitable to the dignity of our birth.

You will say, how shall we manifest our care of our souls? What shall we do for them? I answer, The same we do for our bodies. 1. We feed our bodies; our souls are also to be fed. The food of the soul is knowledge, especially knowledge in the things of God, and the things that concern its eternal peace and happiness; the doctrine of Christianity, the word of God read and preached; 1 Pet. ii. 2. 'As new born babes desire the sincere milk of the word, that ye may grow thereby;' Heb. v. 12. The apostle speaks both of milk and of strong meat. Milk he there calls the principles of the doctrine of Christ; and again, 1 Cor. ii. 3. 'I have fed you with milk and not with meat, for hitherto ye were not able to bear it.' So we see in the apostle's phrase, feeding of the flock, is teaching and instructing of them. Knowledge is the foundation of practice; it is impossible to do God's will before we know it; the word must be received into a honest and good heart, and understood, before any fruit can be brought forth.

Secondly, We heal and cure our bodies, when they are inwardly sick, or outwardly harmed: sin is the sickness of the soul, Matt. ix. 12. 'They that be whole need not a physician, but they that be sick,' saith our Saviour by way of similitude, which he explains in the next verse, 'I am not come to call the righteous, but sinners to repentance.' For the cure of this disease, an humble, serious, hearty repentance is the only physic; not to expiate the guilt of it, but to qualify us to partake of the benefit of that atonement which our Saviour Christ hath made, by the sacrifice of himself, and restore us to the favour of God, which we had forfeited, it being as much as in us lies an undoing again what we have done.

Thirdly, We clothe and adorn our bodies; indeed too much time and too many thoughts we bestow upon that; our souls are also to be clothed with holy and virtuous habits, and adorned with good works, 1 Pet. v. 5. 'Be ye clothed with humility;' and in the same epist. chap. ii. 3. he exhorts women to 'adorn themselves, not with that outward adorning of plaiting the hair, and of wearing gold, &c. but with the ornament of a meek and a quiet spirit, which is in the sight of God of great price:' and in Rev. xix. 8. 'The righteousness of the saints' is called 'fine linen.' And the saints are said to be clothed 'in white raiment.' Matt. xxiii. 11. Works of righteousness and a conversation becoming the gospel, is called a 'wedding garment.' Colos. iii. 10. 'Put on the new man.' And again, 'Put on therefore, as the elect of God, bowels of mercy, meekness,' &c. On the contrary, vicious habits and sinful actions are compared to filthy garments. So Zech. iii. 3. Joshua the high-priest is said to be 'clothed with filthy garments;' which in the next verse are interpreted his iniquities, either personal, or of the people, whom he represented, 'I have caused thy iniquity to pass from thee, and will clothe thee with change of raiment.'

Fourthly, We arm and defend our bodies. And our souls have as much need of armour as they: for the life of a Christian is a continual warfare; and we have potent and vigilant enemies to encounter withal; the devil, the world, and this corrupt flesh we carry about with us. We had need therefore to take to us the Christian panoply, to 'put on the whole armour of God, that we may withstand in the evil day, and having done all may stand; having our loins girt with truth, and having the breastplate of righteousness, and our feet shod with the preparation of the gospel of peace. Above all taking the shield of faith, and for an helmet, the hope of salvation, and the sword of the Spirit, which is the word of God.' Eph. vi. 13, 14, &c.

He that with this Christian armour manfully fights against and repels the temptations and assaults of his spiritual enemies; he that keeps his garments pure, and his conscience void of offence towards God and towards man, shall enjoy perfect peace here, and assurance for ever. Tacitus saith of the Finni, a northern people, that they were ' securi adversus homines, securi adversus Deos.' They need not fear what God or man could do to them, because they were in as bad a condition as would consist with living in the world: they could not be banished into a worse country, nor put into worse circumstances than they were in already. I might say of the man that keeps a good conscience, that he is secure against God and man; not in that sense the Finni were, but secure of any evil befalling him, from either. God can do him no harm, not for want of power, but for want of will, which is regulated by his truth and justice. He is also secure in respect of men, because he is under the protection of the Almighty: and if any there be that would do him harm, they shall either be restrained by the divine providence, or if they be permitted to injure him, it shall tend only to the exercise and improvement of his faith and patience, and enhancing his future reward at that great day; when the Almighty shall dispense *aureolæ* to those champions who have signalized their valour and fidelity by heroic actions, or patient sufferings of unworthy things for his sake. 3. A good conscience not only secures a man from God and men, but from himself too. ' There is no peace to the wicked saith my God,' no inward peace. Such a man is at odds with himself. For the commandments of God being agreeable to the nature of man, and perfectly comformable to the dictates of right reason ; man's judgment gives sentence with the divine law, and condemns him when he violates any of them ; and so the sinner becomes an *Heautontimorumenos,* a tormentor of himself. ' Prima est hæc ultio, quod se judice nemo nocens absolvitur.' No guilty person is absolved at his own tribunal, himself being judge.

Neither let any profligate person, who hath bidden defiance to his conscience, and is at war with himself, think to take sanctuary in atheism, and because it imports him highly there should be no God, stoutly deny that there is any. For first, Supposing that the existence of a Deity were not demonstrably or infallibly proved (as it most certainly is), yet he cannot be sure of the contrary, that there is none. ' For no man can be sure a pure negative, namely, that such a thing is not, unless he will either pretend to have a certain knowledge of all things that are or may be, than which nothing can be more monstrously and ridiculously arrogant; or else unless he be sure that the being of what he denies, doth imply a contradiction ; for which there is not the least colour in this case. The true notion of God consisting in this,-that he is a being of all possible perfection.' That I may borrow my Lord Bishop of Chester's words in his Discourse of Natural Religion, p. 94.

Now if he be not sure there is no Deity, he cannot be without some suspicion and fear that there may be one.

Secondly, ' If there should be a Deity, so holy, and just, and powerful as is supposed, what vengeance and indignation may such vile miscreants and rebels expect, who have made it their business to banish Him out of the world, who is the great Creator and Governor of it, to undermine his being; and eradicate all notions of him out of their own and other men's minds; to provoke his creatures and vassals to a contempt of him, a slighting of his fear and worship, as being such imaginary chimeras, as are fit only to keep fools in awe. Certainly as this is the highest provocation that any man can be guilty of, so shall it be punished with the sorest vengeance.'

Now a slender suspicion of the existence of a Being, the denial whereof is of so sad consequence, must needs disturb the atheist's thoughts, and fill him with fears, and qualify and allay all his pleasures and enjoyments, and render him miserable even in this life.

'But on the other side, he that believes and owns a God; if there should be none, is in no danger of any bad consequent. For all the inconvenience of this belief will be, that he may be hereby occasioned to tie himself up to some needless restraints during this short time of his life, wherein notwithstanding there is, as to the present, much peace, quiet, and safety; and, as to the future, his error shall die with him, there being none to call him to an account for his mistake.' Thus far the bishop. To which I shall add, that he not only suffers no damage, but reaps a considerable benefit from this mistake; for during this life he enjoys a pleasant dream or fancy of a future blessed estate, with the thoughts and expectation whereof he solaces himself, and agreeably entertains his time; and is in no danger of being ever awakened out of it, and convinced of his error and folly, death making a full end of him.

THE END.

Printed by J. F. Dove, St. John's Square.